Nifty E-Z Guide
EchoLink™ Ope

By Bernie Lafreniere, N6FN

**Another book in the
Nifty! Ham Accessories
E-Z Guide Series**

www.niftyaccessories.com

Copyright

Copyright © 2011 by Nifty Ham Accessories / Bernard Lafreniere – N6FN. All rights reserved, no part of this book or portions thereof may be reproduced in any form or by any means, electronic or mechanical, including photocopying, recording, or by any other means, without permission in writing from the publisher.

Disclaimer and Limitation of Liability

While every effort has been made to make this publication as accurate as possible, Nifty! Ham Accessories and the author assume no liability for the contents regarding safety or damage to equipment, violations of FCC rules and do not guarantee the accuracy herein.

EchoLink™ is a registered trademark of Synergenics, LLC.

Contents

About This Guide 1

Special Thanks 2

Introduction 3

Chapter 1: Internet Voice Communications 4
   Voice Over Internet History 4
   Digital Sound Conversion.. 4
   Audio Compression 6
   Routing Audio via the Internet 6
   PC Based VoIP Data Transmission 7

Chapter 2: Internet Repeater Linking 9
   Early Experiments.. 9
   But do we have a System? 10

Chapter 3: The EchoLink System 13
   EchoLink Node Types 14
   Individual User Nodes 14
   RF Simplex Link Nodes 15
   RF Repeater Nodes 16
   Conference Servers 17
   Centralized Addressing and Validation Servers 19

Chapter 4: Installing EchoLink Software 21
   System Requirements 21
   Downloading the Software 21
   EchoLink for Linux and Apple's Mac OS X 22
   Software Installation and Basic Configuration 22
   Firewall / Router Tester 23
   Firewall / Router Problems 24
   Configuring Firewalls and Routers... 25
   Windows 7 and Vista Firewall Issues... 27

Call Sign Validation / Authentication 27
Verifying Your Node Has Been Added 28
Setting Audio Parameters and Microphone Type 30
Accessing the Test Server and Adjusting Volume Levels 31
Selecting the Transmit Key and its Operating Mode 33
Customizing Your Station Information 35
Verifying How Your Name Appears 36

Chapter 5: Operating EchoLink 37
Initiating a QSO Via Computer 37
Connecting via the Station List 38
Connecting to Conference Servers 39
Operating Procedures 40
Direct Computer-to-Computer Contacts 42
Chat / Texting 43
Disconnecting 43
Keeping Track of Favorites 44
Setting Station Alarms 44
Low-Bandwidth Indicator 45
Call Sign Log 46
DTMF Keypad Commands 47
Initiating a QSO with a Radio 48
Initiating a QSO via a Wireless Hot Spot 51
EchoLink for iPod Touch, iPad and iPhone 54
EchoLink for Android Devices 54
EchoLink Nets 55
Useful Web Pages 55

Chapter 6: Establishing an RF Node 57
Selecting Sysop Mode 57
Interfacing a PC to a Transceiver 58
Connecting Sound Card Audio to the Transceiver 60
Keying your Transceiver 60
Establishing a Repeater Node 61
PC to Transceiver Interfaces 61
EchoLink Specific Interface Features 62

Chapter 7: Commercial PC Interfaces 64

Chapter 7, Section 1: RIGblaster Plus II Interface 64

Distinctive Features of the RIGblaster Plus II 65
RIGblaster Plus II Interface Cables 66
RIGblaster Plus II Functional Block Description 67
RIGblaster Plus II Installation and Set Up..................................... 68
Sysop Settings for use with the RIGblaster Plus II....................... 69
Setting Received Audio Drive Level from the Transceiver........... 70
Setting EchoLink's Received Audio VOX Threshold................... 71
Setting Audio Drive Level to the Link Transceiver...................... 73

Chapter 7, Section 2: WB2REM/G4CDY ULI Interface 76
Distinctive Features of the WB2REM / G4CDY ULI Interface 76
ULI Interface Cables and Power Source... 78
ULI Functional Block Description 79
WB2REM / G4CDY ULI Installation and Set Up......................... 81
Sysop Settings for use with the ULI 82
Setting Received Audio Drive Level from the Transceiver........... 85
Setting EchoLink's Received Audio VOX Level 86
Setting Audio Drive Level to the Link Transceiver...................... 87

Chapter 7, Section 3: SignaLink USB Interface............................. 90
Distinctive Features of the SignaLink USB Interface.................... 90
SignaLink USB Functional Block Description............................... 92
SignaLink USB Installation and Set Up 94
Installing the Jumpers 94
Installing the Window's Audio Codec Driver 95
Sound Card Selection... 95
Sysop Settings for the SignaLink USB 97
Setting Received Audio Drive Level from the Transceiver........... 98
Setting EchoLink's Received Audio VOX Threshold................... 99
Setting SignaLink's VOX and the Transmit Drive Level............ 100

Chapter 7, Section 4: Other PC to Radio Interfaces 104

Chapter 8: Link Radio Requirements 106
Output Power Rating 106
CTCSS, Tone Squelch 106
Carrier Operated Switch, COS.. 107
Received Audio Characteristics... 108
Power Supply Requirements.. 109
Amateur Transceivers 109
Auxiliary Data Connectors on Amateur Transceivers.................. 110

Surplus Commercial Transceivers 111
Commercial Transceiver Yahoo Groups.. 112

Chapter 9: Sysop Mode Setup 113
RX Ctrl Tab... 113
TX Ctrl Tab.. 115
DTMF Tab 115
Ident Tab 118
Options Tab... 119
Signals Tab.. 121
Remt Tab... 122
RF Info Tab... 123
Preferences Connections Tab... 124
Remote Control Web Server 126

Chapter 10: Eliminating IDs and Tones............................... 128
Node Co-location at the Repeater Site 128
Remote EchoLink Repeater Node Location 129
Using CTCSS.. 130
Passing CTCSS through the Repeater.. 131
Preventing Simplex Node Interference 132

Chapter 11: FCC Rules and VoIP 133
Part 97 Definitions 134
Single-User Computer or Transceiver operation 134
Establishing Your Own RF Link 135
Remote Control Methods 135
Control Operators and the Control Function.............................. 136
Link Transceiver Call Sign ID 137
Auxiliary Control Station Frequencies... 137
FCC Part 97 Rules and Regulations.. 138

Chapter 12: Simplex Link Frequencies 139

Appendix A: PC Serial Port Pinout...................................... 140

Appendix B: Entering Call Signs via DTMF........................ 141

Appendix C: **Radio Setup Guides** **142**

About This Guide

Today the two most popular amateur radio voice over Internet systems are undoubtedly EchoLink and IRLP (Internet Radio Linking Project). While this book is primarily about EchoLink, both EchoLink and IRLP have the goal of interconnecting amateur radio stations and repeaters using VoIP, Voice Over Internet Protocol technology.

While EchoLink and IRLP in many ways are similar in concept, they are substantially different in implementation and operation. It would be difficult to say that one is better than the other – they are just different. Once the operating characteristics of both systems are understood, many hams discover they have a preference for one or the other. Akin to people's preferences in their choice of radios, a decision is often influenced by personal preference. A person's choice might be influenced by the ready availability of either an EchoLink or IRLP capable repeater, either in their local area or in areas they wish to make contact with.

Using easy-to-understand language and illustrations, this guide describes how to access and use EchoLink-equipped radio links, repeaters and conference servers. Several chapters provide guidance for those wishing to set up their own RF based EchoLink simplex and repeater nodes. The book is light on the technical aspects of the underlying Internet data packet transporting mechanisms, concentrating instead on the practical issues of understanding EchoLink's capabilities, getting things configured and making contacts.

For those wishing to use their computers or make use of existing EchoLink repeater nodes for making contacts, you only need to review Chapters 1 through 5. Those interested in establishing their own RF-based simplex or repeater nodes will benefit by reading Chapters 6 through 12.

Lets get started!

Special Thanks

I wish to thank all those that helped in the creation of this book. A special thanks goes to the reviewers that took of their valuable time to proofread and offer suggestions that materially improved the manuscript: Norbert, KJ6ZD, developer of EI-151 and EI-160 EchoLink interfaces; John, KT6E and Charlie, NN3V both active DXers and long time friends; and Michelle, W5NYV, editor of the Palomar Amateur Radio Club radio newsletter.

I'm especially grateful to Paul Williamson, KB5MU, a software engineer, who was involved with Qualcom's early CDMA cell phone development and who has contributed significantly to amateur radio satellite tracking software. He was of immeasurable assistance in applying a magnifying glass to the text, asking thought provoking questions and making numerous suggestions.

Jim, WB2REM was helpful in answering questions about the operation and set up of the ULI interface. The Palomar Amateur Radio Club helped out by permitting me to install a repeater node on one of their repeaters in Southern California. Art, KC6UQH, who modified one of the Palomar Amateur Radio Club repeater controllers to better support EchoLink. And my friend Dave, N7NKK, was instrumental in securing permission for hosting a repeater node on the Mount Dutton repeater in Southern Utah.

I also wish to thank Jonathan Taylor, K1RFD, the creator of EchoLink, for reviewing the final manuscript and pointing out several things that needed mentioning that I had missed.

Introduction

EchoLink can now be considered a mature technology. Even though the system was designed ten or more years ago, the popularity for this method of long-distance communication remains undiminished. The EchoLink network of simplex links, repeaters, conference servers and individual users has continued to grow, and today there are hundreds of thousands of registered users in more than 162 different countries. At any given time there may be five thousand or more repeaters and users on line.

The appeal of using EchoLink for accessing repeaters through the Internet can be attributed to the facts that it is: free, easy to use, and perhaps most importantly because it opens up long range (DX like) communications to Technician class license holders and others that may not be able to set up traditional HF stations with large antennas.

Since the Internet does not rely on the vagaries of HF radio wave propagation characteristics, EchoLink can provide 24/7 communications to anywhere the Internet is available. Even for those with HF capability, EchoLink is a useful and interesting adjunct to anyone's shack. Its potential as an alternative method of emergency communications just adds to the reasons why this mode of operation is so popular.

Chapter 1: **Internet Voice Communications**

Hams have a long history of applying new technologies to amateur radio communications. The Internet has been one of the recent technological advances that innovative hams have been experimenting with in their quest for improved ham radio communications. Transporting voice contacts over the Internet, commonly referred to as VoIP (Voice Over Internet Protocol), has been one of the technologies hams have been quick to take advantage of. EchoLink and IRLP are two examples of how VoIP technology has been applied to accessing or linking remote VHF/UHF repeaters.

Voice Over Internet History

Internet Phone, one of the first widely available VoIP computer applications became available to users of the Internet in the 1990's. This and similar applications allowed you to attach a microphone to your PC and make free phone calls to other *Internet Phone* equipped computer users. While this technology was not entirely trouble free, it was sufficiently intriguing that it inspired a number of hams to find ways of applying the technology to ham radio.

Since the basics of how *Internet Phone* worked still form the basis for the way voice-over-Internet applications work today, it is worthwhile to take a moment to review how it was done.

Digital Sound Conversion

Back in the 1990's PCs were equipped with a device known as a sound card. Today it's most likely included within one of the chips located on the motherboard instead of a separate card plugged into the computer. In either case, it is the system component responsible for converting sound waves into a stream of digital numbers when recording, and vice versa, converting a stream of digital numbers back to a sound wave when playing back recorded or received audio.

An electronic device that converts an analog value into a digital number is called an analog-to-digital converter, or simply an A-to-D converter. If one performs samples (A-to-D conversions) fast enough, a varying analog signal such as a voice waveform can be converted into a stream of numbers which represents the voltage amplitude of the waveform at each of the sampled points.

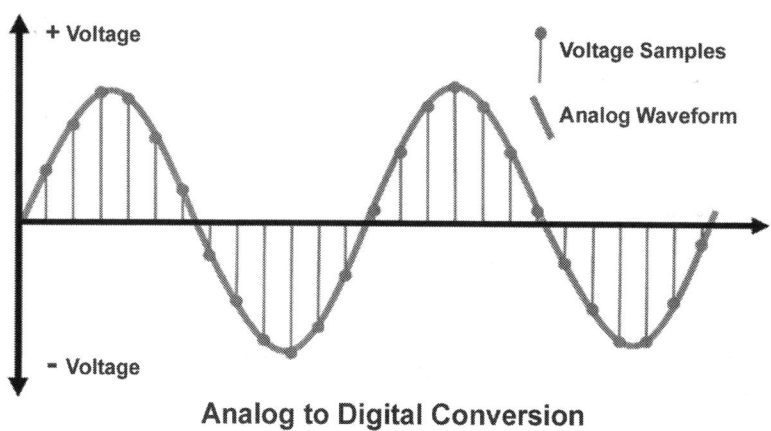

Analog to Digital Conversion

As can be seen in the above diagram, each of the points along the waveform can be represented by a numerical value indicating the measured voltage at that point. Streams of these numbers, representing the original sound waves can be recorded for later playback.

During playback, when a stream of these numbers is passed through a digital-to-analog converter (D-to-A converter), the encoding is reversed and an approximation of the original waveform is reproduced.

With a little bit of filtering, the resulting reconstituted audio wave can be made to sound as good as we need it to be. For high quality stereo (think CD quality) a relatively high rate of 44,100 samples per second is typically used. This high rate of sampling provides a bandwidth approaching 22 kHz, beyond the range of hearing for most humans. This high sampling rate combined with a high degree of resolution on each sample, 16 bits, allows the input waveform voltage to be converted quite precisely with a numerical value that can range between +/- 32,000. The resulting audio quality is superb, but the drawback is the massive amount of data that it generates.

To reduce data rate requirements for telephony and radio communication applications, it has long been established that an audio bandwidth of about 3 kHz is adequate for voice communication. Consequently, for communication quality digital conversion, standard practice has been to use a waveform sample rate of 8,000 samples per second, which yields a 4,000 Hz bandwidth. To further reduce the amount of data that needs to be transmitted, the numerical precision used to measure each of the waveform samples is reduced to 8 bits, which can represent a numerical range of +/-128.

Audio Compression

In addition to reducing the sampling rate and numerical conversion precision, various data compression techniques can also be employed to further reduce the amount of data that needs to be transmitted. EchoLink uses a form of compression known as GSM, *Global System for Mobile Communication.* Popularly used for cell phone communication, GSM has been optimized for narrow bandwidth connections, allowing EchoLink to work with the data rate limitations of conventional phone lines.

While it is true that all this data reduction and compression tends to reduce the overall quality of the audio, it produces audio that compares favorably with the processing techniques used in telephony and radio communication services we are accustomed to. While it's not stereo CD quality, it's more than adequate for communication purposes.

Routing Audio via the Internet

Once an audio stream has been sampled and compressed, it still needs to be routed via the Internet to its intended destination. To accomplish this, the designers of a VoIP system need to convert the digital audio data stream into data packets compatible with the Internet protocol for addressing and routing information over the Internet. This protocol allows the packets to make their way through a maze of public and private telecommunication networks and other pieces of equipment en route to their intended destination.

At the receiving end, these data packets need to be decoded, buffered and reassembled into the proper order and finally converted back into a fair reproduction of the original analog sound wave suitable for outputting to a headset or speaker.

PC Based VoIP Data Transmission

By making use of a PC that is connected to the Internet, application software can convert a microphone's audio input into a stream of digital numbers, which when sent over the Internet can be received by a similarly equipped computer and converted back into a good approximation of the original audio.

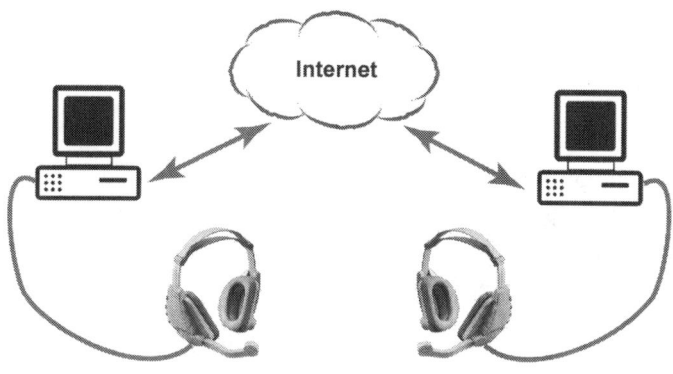

PC-to-PC VoIP Communication

A PC software application handles recording, formatting and transmitting the audio stream to the distant computer. The details for converting the digital data to packets and sending them over the Internet can be quite complex and is a credit to the skills of the engineers and programmers who have created these systems.

While the initial *Internet Phone* and similar early applications were not without problems, newer VoIP applications have made good use of faster computers, better operating systems, and higher Internet data transmission rates to dramatically improve upon those early VoIP implementations.

In recent times this technology has been improved to the point where it has moved from the realm of experimentation to being available to the general public in a number of commercial products. Vonage, Skype and other companies now provide both low-cost and free Internet telephone services. In the realm of ham radio, EchoLink and *IRLP* are popular VoIP choices.

Note: D-STAR, introduced by Icom, is also quite popular. It provides an alternative and interesting method of linking remote nodes and repeaters, but its architecture is quite different from traditional VoIP systems.

EchoLink and IRLP both make use of conventional FM repeaters and transceivers, transmitting audio modulated RF. After an FM receiver has recovered the audio signal, conversion of the audio into digital packets for transport over the Internet occurs in a PC. Digital technology is not used over the amateur band RF links.

D-STAR on the other hand is true digital radio technology. Expressly designed digital-capable radios and repeaters transmit and receive digital data packets directly – not audio modulated RF. *Internal to the radio*, voice is converted to a digital format using an electronic chip called a CODEC, which encodes and decodes audio signals in the AMBE (Advanced Multi-Band Excitation) digital format, a proprietary speech coding standard developed by Digital Voice Systems, Inc. Consequently when a D-STAR capable radio is operating in D-STAR mode, it transmits packets of digital information. These packets not only contain the audio stream, but also carry embedded repeater node routing data, call signs, error checking and other information.

Chapter 2: Internet Repeater Linking

Once enterprising hams had experimented with *Internet Phone* and similar VoIP applications, it didn't take long to realize that by connecting a transceiver's speaker output to the sound card's microphone input and by connecting the transceiver's microphone input to the sound card's audio output jack, that it should be possible to link two transceivers together via Internet connected computers.

Early Experiments

While there were audio level differences to overcome and other shortcomings, early experiments did exactly this: interconnecting two remote VHF/UHF FM transceivers via the Internet.

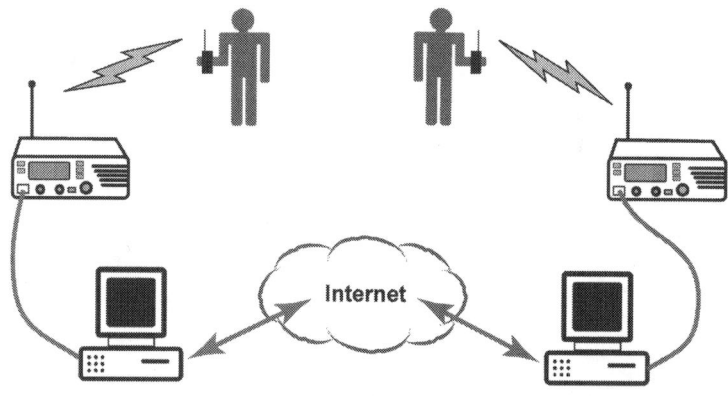

Simplex Link-Transceiver to PC VOIP Communication

Operating simplex, anyone transmitting within range of the link transceivers could now be heard at a distant person's receiver. The net result is half-duplex communication between remote locations using VoIP as the transport mechanism.

Other than a small amount of delay in the audio and perhaps occasional short breaks in the audio stream, the effect is essentially like having a normal simplex conversation. Only one person at a time can speak. It does not matter if the simplex frequencies are different, or even cross-banded.

Once basic simplex operation was achieved, it was a fairly small step to reprogram the link transceiver for communication with a standard VHF or UHF FM repeater.

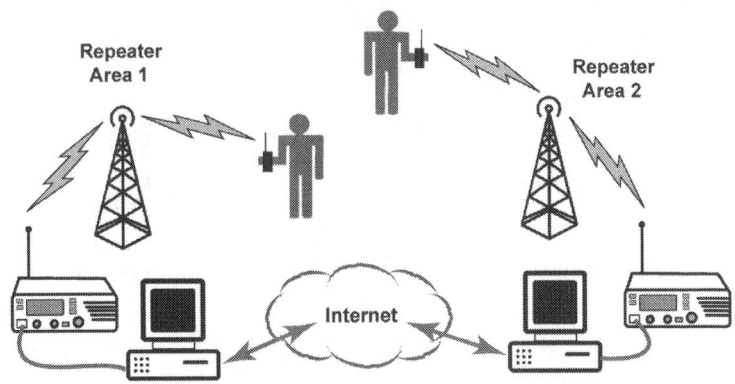

Link-Transceiver to Repeater VOIP Communication

When both ends of the link have their transceivers tuned to repeaters, anyone within range of either repeater can hear both sides of a conversation in progress. The person talking does not hear his own transmission of course, but others tuned to the repeater he is using will hear him on the repeater output as usual. Users listening to the remote repeater will also hear the distant person's transmission.

But do we have a System?

Cobbling together some software and hardware to link a couple of repeaters together to prove the concept is a far cry from having an easily used application with sufficient features to give a large number of users the capability of linking with an arbitrary number of remotely located repeaters.

Since early experimentation clearly demonstrated that linking repeaters together via the Internet was possible, the next step was to develop a software application allowing individual hams to set up and access a network of linkable repeaters. It was to this end that the developers of EchoLink and IRLP applied their creativity, engineering skills and perhaps most of all, perseverance.

You might be thinking: "Heck, if we can link repeaters together, what else is there to do?" And in one sense you would be right. If the vision was to interconnect a couple of repeaters, say one in San Francisco and another in New York, perhaps not a whole lot more would need to be done. Hams at either end of the link could contact each other, almost as if they were operating on the same repeater. We could even throw in a couple more repeaters and they could all be linked together. The effect would be somewhat similar in operation to existing RF linked repeater systems.

But instead, what if the vision is to allow individual hams the ability to link to distant repeaters of their choice, when they want? Imagine an individual ham, operating on his local repeater, being able to transmit a few DTMF tones using his transceiver's keypad and bringing up a repeater in another state or another country? Oh wow! That is a different concept all together.

If we want to do that, a lot of issues and problems come up that are beyond the technical ones involved with converting sound to a digital format and transporting it over the Internet:

- Since all the stations need to be compatible with each other, what kind of software needs to be written?
- Who maintains it?
- How will the software be distributed?
- How will new versions of software be distributed to remotely located repeaters?
- How will users find out what stations are currently on line?
- How are stations addressed?
- Who administers the station addressing process?
- How will stations access remote stations?
- What do we do if a station is already in a conversation?
- How is the system made robust enough so that no single point of failure brings down the system?
- Since FCC and foreign country regulations are involved with the RF portions of the link, what needs to be done to stay legal?

- If the network is to be global, how do we accommodate the regulatory rules of foreign countries?
- What is the hardware and software architecture that operators of a linked repeater need to implement?
- How are repeaters and individual stations connected to the network?
- How are multiple interconnected stations managed?
- What computer operating systems need to be accommodated?
- How do we ensure that access is limited to licensed hams?
- What features need to be implemented in the user interface?
- How do we make it easy to use?
- How do we link multiple repeaters together for wide-area nets or emergency communications?
- What operating policies need to be implemented?
- What software features are required for system administrators to manage their systems and stay compatible with the network?

Fortunately for ham radio a number of enterprising and talented amateurs took on the challenge of solving these problems and many more.

Building upon a successful application known as *iLINK*, developed by Graeme Barnes, M0CSH, Jonathan Taylor, K1RFD, produced an early version of the EchoLink system in 2002. It also underwent a series of improvements to bring us the robust system that we currently have. Today the EchoLink system is successfully providing users with a reliable network that spans the world. Thousands of hams enjoy using this great system. Truly a magnificent achievement!

Chapter 3: **The EchoLink System**

The EchoLink system is interesting in that users can link to remote nodes either by using EchoLink software on their computer or by keying DTMF commands directly from their transceiver. Using DTMF commands from a transceiver requires communication with either an EchoLink equipped repeater or a simplex link.

If you don't have an EchoLink equipped repeater or simplex linked node available in your area, you can still access distant nodes by downloading and using the EchoLink software on your computer. This is an important point: even if you live hundreds of miles from a repeater, you can still make contacts on distant repeaters using your computer!

As part of setting up the software there is a process by which you submit your call sign and other information so that the administrators of the EchoLink system can validate that you are a currently licensed amateur radio operator. Upon being validated you will be issued a node number and your EchoLink software will become fully operational, allowing you to see in real-time stations that are online and their current operating status. (In actual practice there may be a delay of a few minutes before status changes appear.)

When within operating range of an EchoLink equipped repeater or simplex linked node, you can use a transceiver to key a string of DTMF command tones to connect to distant nodes. But you need to know the DTMF command codes and the number (address) of the node that you wish to contact. The list of DTMF command codes and currently active node numbers can be found by using your EchoLink software or by looking them up on the repeater sponsor's club or individual's web page.

Node numbers and the on-line / off-line status of simplex link and repeater nodes can also be found on the *EchoLink Link Status* web page, by selecting the *Link Status* sidebar button on the EchoLink website's home page. **http://www.echolink.org/**

System operators of individual RF-connected simplex links and repeaters have the option of either using the standard EchoLink command codes, or modifying them to suit their own purposes. Consequently, it's best to contact the node's sponsor to find out what their procedures and use policies are. Often these are available on the node sponsor's web page.

Contacts can be made from your computer or a DTMF equipped transceiver to four different node types: another user's computer, a simplex link, a repeater, or a Conference server, all of which are described below.

EchoLink Node Types

The EchoLink system supports four different types of user accessible Internet nodes:
- Individual Users on computers   N6XXX
- RF Simplex Links   N6XXX-L
- Repeaters   N6XXX-R
- Conference Servers   *N6XXX* or *NAME*

As can be seen above, each type of node has a uniquely formatted call sign or name. Individual computer users use their basic call sign. Other types of nodes have a suffix appended to the call sign, indicating the type of node: **-L** for a simplex link, **-R** for a repeater, or in the case of a conference server, a call sign or server name bracketed on both sides with an asterisk. In addition to the call sign, which is generally the call sign of the sponsoring individual or club, each node is assigned a unique node number, which can vary from four to six digits.

Individual User Nodes

Once individual users have installed and configured the EchoLink software, whenever they are connected to the Internet and running the EchoLink program, they are visible as an active EchoLink node capable of initiating *or receiving* node connections. A few moments after starting a node's EchoLink software, its call sign will show up as being on-line in the application's station list of every user on the system.

Individual users are identified by the node number assigned during validation and by their call sign, with no suffix attached.

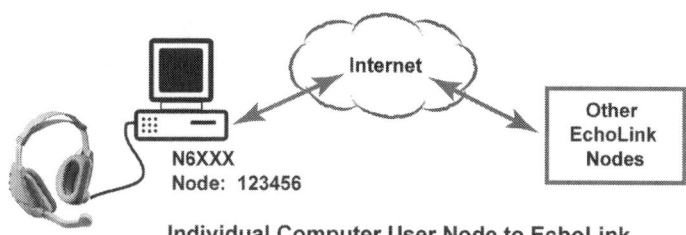

Individual Computer User Node to EchoLink

Important: If, like many users, you do not intend to set up an EchoLink repeater or RF simplex node, and will be operating EchoLink as a single user via your own computer, or via RF to an already EchoLink equipped repeater, or through someone's simplex node, select single **USER** mode when installing the EchoLink software. In **USER** mode, set up is simplified. By not selecting EchoLink's **SYSOP** mode, you won't need to be involved with the complexities of interfacing a transceiver to a computer and the intricacies of configuring an EchoLink RF based node's operation.

RF Simplex Link Nodes

If desired, individual users can select **SYSOP** mode and create an RF simplex link by interfacing a transceiver, tuned to a clear simplex frequency, to the microphone and speaker jacks of their computer's sound card. Even if you do not create your own simplex link, you can still access stations by using other people's simplex links.

Stations within range of a simplex transceiver can access the EchoLink network, establishing node connections as desired by transmitting DTMF command strings. Once the link has been established, other stations within range of the simplex transceiver will also be able to communicate with the distant node.

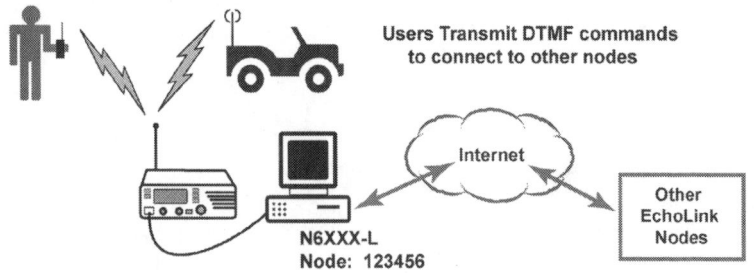

RF Simplex Link Node

Simplex links are identified by the user's call sign followed by the –L suffix, and by a unique node number assigned during validation of the –L call sign. Simplex link owners are often happy to let other hams use the link. In fact, most of them encourage others making use of the capability they have created.

RF Repeater Nodes

Repeater Links usually are created by interfacing a transceiver tuned to an existing repeater to the microphone and speaker jacks of a computer's sound card. In this way a transceiver and Internet connected computer can be used to link a normal FM repeater to the EchoLink network. For stability and 24/7 operations, the computer is usually dedicated to the EchoLink task and is not used for other applications.

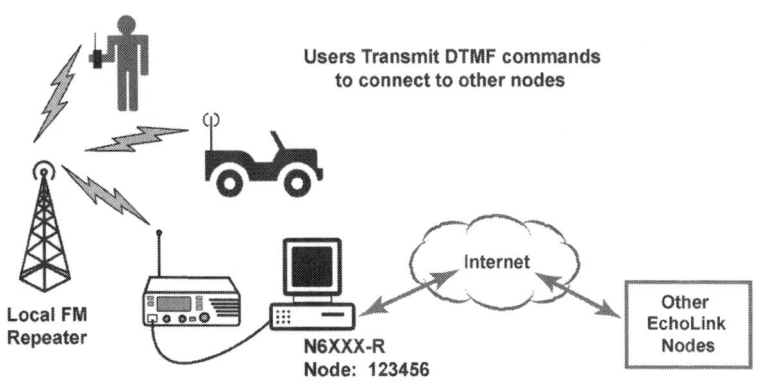

RF Repeater Node

Repeater nodes are identified by their node number, and by the sponsor or club's call sign followed by the –R suffix. Repeater node owners are often happy to let other hams use the link, encouraging them to make use of the repeater's EchoLink capability. On the other hand, some repeaters are not "open" repeaters and may be restricted to those providing support for the system or for some other reason.

Stations within range of the repeater can access the EchoLink network, establishing node connections as desired, by transmitting the necessary DTMF command strings. Once the link has been established, other stations within range of the same repeater are also be able to communicate with the distant node.

The advantage of a repeater node is that anyone within range of a repeater's coverage area can access the EchoLink network with their radio, establishing connections as desired. In most cases, this will be a wider footprint than a Simplex link node. A potential difficulty with this setup, is that the system administrators (sysops) who install the system need to ensure that the repeater's ID's, courtesy tones and squelch tails are not transmitted over the Internet. (More on eliminating repeater generated noises in later chapters.)

If Internet service is available at the repeater site, it's also possible to directly connect the repeater to the PC's sound card, eliminating the need for a link transceiver. A significant advantage in being co-located with the repeater is that it generally makes it easier to eliminate unwanted repeater signals from being transmitted over the EchoLink network. Making use of a carrier signal detected at the receiver as the condition for enabling transmissions to the Internet automatically eliminates repeater ID's, courtesy tones and squelch tails, which are generally only present on a repeater's transmit side.

Conference Servers

EchoLink nodes configured as conference servers connect multiple users together, as if they were connected to a single conventional repeater. All types of EchoLink nodes can connect to conference servers: repeaters, individual users, simplex links and if enabled, other conference servers. An Internet connected conference server receives audio packets from a transmitting station's local node and re-transmits them to all connected nodes. Participants in a conference can all hear each other, and typically take turns transmitting in round-table fashion. In this way many EchoLink nodes can be tied together, which can be especially useful for wide-area nets or emergency communications.

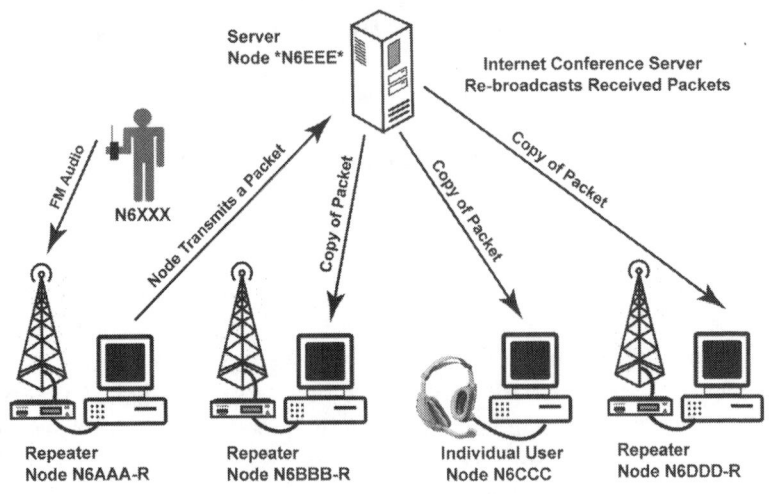

Conference Server Operation

Conference servers are identified by a node number, and by either the sponsor or club's call sign bounded on both sides with an asterisk *N6XXX*, or by an arbitrary textual name, also bound on both sides by an asterisk *NAME*.

Since EchoLink uses an Internet protocol designed to direct traffic from point to point, for multiple nodes to be able to hear a transmitting node, a conference server is used to re-transmit the signal to all the other participants of the conference. To do this, the conference server maintains a list of all of the participants in the conference, taking the audio stream received from one participant and distributing it out to all the others. For instance, if there were 20 participants in a conference, the server would be receiving data from one participant while simultaneously re-transmitting it to the other 19 participants. Each of those 19 participants is sent their own copy of the audio stream. Because audio streams need to be continuously received without interruption, conference servers require high-bandwidth Internet connections.

In addition to dedicated Internet conference servers, the PC-resident EchoLink application supports conferencing in both single-user and sysop modes. If conferencing is enabled, the number of participants should be limited to what can be supported by the speed of the node's Internet connection. For example, conferencing is not recommended on dial-up lines, while typically a cable modem can support about seven users. Higher data rate Internet connections can support even more participating stations.

Centralized Addressing and Validation Servers

Users don't directly access addressing or validation servers; they operate hidden in the background. Somewhat like traffic cops checking licenses, they exist to keep the EchoLink Internet-highway running smoothly and to keep out the bad guys (un-validated users).

EchoLink uses a set of centralized addressing servers to handle a variety of system administration responsibilities. Several of these servers are used to provide redundancy in case one or more should fail or be unreachable for some reason. In addition, the servers use a load balancing mechanism to keep things from backing up during periods of high demand and for sharing the load if one or more should fail.

Addressing servers maintain a list of all registered EchoLink nodes, keeping track of which ones are logged in, if they are busy or not, and their current Internet address.

The addressing servers communicate with each other, making sure they are all up-to-date with new stations coming on-line and with stations that are currently logged-in. In addition to keeping track of all these addresses and the list of currently active stations, the servers are used to validate that new users are currently licensed hams and that they are in fact who they say they are and not pirating someone else's call sign. Before being able to use the EchoLink software and establish a node, all users must first be validated by EchoLink's administrators.

Once they have been validated, each registered user has a password associated with his call sign and node number. The validation information, call sign and password data is stored on yet another server, called the validation server. A shorter version of this data is also kept on the individual addressing servers.

Every time a user starts the EchoLink software, their password and call sign are sent to one of the addressing servers and are verified before allowing access to the system. To make sure the password is secure while being sent over the Internet, it is encrypted using a highly secure technique called "public-key cryptography".

While the above description of the addressing and validation servers is somewhat simplified, it does bring home the point that there is a lot of centralized computer processing going on in the background to keep the EchoLink network running smoothly. Not to mention the expense of maintaining these computers and the required Internet bandwidth resources. We should be thankful to the many volunteers and corporate entities that provide the necessary hardware and Internet bandwidth so that we ham operators can have free access to worldwide communications over the Internet.

Chapter 4: Installing EchoLink Software

System Requirements

EchoLink is designed to work on PCs equipped with any of Microsoft's operating Systems, Windows 95, 98, Me, NT, XP, 2000, Vista and now Windows 7. Almost any PC should be fast enough to run the application. More recent operating systems, though, might be less susceptible to system crashes.

There are a few potential problems that might arise with Win95, though, see **Using EchoLink with Windows 95**, in the **Installation** section of the FAQ's at **http://www.echolink.org/faq_installation.htm**.

Your system will need a sound card (or chip) and about 5 megabytes of free disk space, trivial by today's gigabyte standards. In addition you will need an Internet connection. The application will work on dial-ups capable of 24 kbps or faster. Of course, having either DSL or a cable modem is better, and required if you want to support conferencing multiple users on your system.

Downloading the Software

You can download the software from the EchoLink web page: **http://www.echolink.org/**

Once on the EchoLink web page, select the **Download** link on the left side of the page and then enter your call sign and email address on the page that comes up. Pressing **Submit** brings you to the page where you can download the software.

In addition to downloading the software you can also download and save a copy of Jonathan's *EchoLink Users Guide*. This user guide is essentially the same information found in the EchoLink software's help files – which by help file standards are quite well written. If you are like me, you may want to print sections of the guide needed during set up of the software. It's handy to have a printout to refer to rather than having to switch back and forth between the actual application and the help file screens while getting things set up.

Incidentally, if you are an iPod, iPad or iPhone user you can also download compatible EchoLink software for free by clicking on the **AppStore** link located on the EchoLink home page. **http://www.echolink.org/**. A similar app is also available for Android Smart Phone users.

EchoLink for Linux and Apple's Mac OS X

There are even EchoLink compatible applications available for those that prefer the Linux or Apple operating systems. Since Jonathan Taylor, K1RFD, author of the original EchoLink Windows program, did not write these programs, the appearance of the user interface and features included in the applications are significantly different than the EchoLink application described in this book.

SvxLink, is a Linux program that can be used to connect to other nodes on the EchoLink network. The program can be made to work with a variety of Linux distributions. For more information and to download the program visit:
http://sourceforge.net/apps/trac/svxlink

EchoMac, is a Macintosh OS X client program that allows single user connection to other EchoLink nodes, including conference servers. For more information and to download the program visit:
http://echomac.sourceforge.net/

Software Installation and Basic Configuration

After downloading and saving the EchoLink install file to a convenient place, locate and double click on the file to run the EchoLink installer program. Unless the version has been updated, it is a file named **EchoLinkSetup_2_0_908.exe**.

Note: If you have difficulty installing the software with Windows 2000 or earlier operating systems, you may need to update the Windows operating system's installation software. To do this, return to the web page that you installed the EchoLink software from and run either the "Windows Installer for 95/98/Me", or the "Windows Installer for NT and 2000" utilities.

When the EchoLink software runs for the first time, it comes up with a **Setup Wizard** that helps you configure the program for your system. If you don't intend to set up a transceiver as an RF repeater link or as a simplex link select **Computer User**; otherwise select **Sysop**. If you are unsure, start off with Computer User, you can always change this later. *The Computer User configuration is simpler to set up and is the mode the remainder of this chapter assumes you have selected.* The Sysop settings are covered in subsequent chapters.

Continue with the **Setup Wizard** entering your **Callsign** exactly the way you want it to be registered on EchoLink. If you have registered before, use the same call sign you previously registered with. If you are registering as a **Computer User**, use your FCC issued call sign with no suffixes added to it. If you are registering as a **Sysop**, append your call sign with either **–R** to indicate a repeater node, or with **–L** to indicate a simplex link node.

Note: If you have previously registered, and are re-installing the software or installing the software on another computer, your registered call sign and password will be recognized, allowing the new software installation to proceed.

Create a **Password**, making a note of what it is in case you need to re-install the software or generate additional call signs for operating your own Simplex or Repeater nodes. If you have registered before, use your original password. Enter your **First Name** and station **Location** (or station description) the way you want them to appear on other stations' screens.

If you have registered before and have forgotten your password, you can request that a password reminder be sent to you via email at **http://www.echolink.org/validation/pw_email.jsp**. If your e-mail address has changed since you last registered, your call sign will need to be re-validated.

Firewall / Router Tester

The next step in the installation process is to run the **Firewall/Router Test** utility, which pops up. If the test runs without error for both TCP and UDP connectivity, you are home free and can continue with the set up process. And if you like, you can skip the following firewall / router problem paragraphs, continuing with the following **Call Sign Validation / Authentication** section.

If the above test failed, and you have been successfully using the Internet for other applications; the TCP connectivity part of the test is most likely passing. If the test failed, your computer or router's firewall may not be allowing outbound traffic over TCP port 5200. For more information and suggested solutions to this problem, refer to the **Firewall Solutions** section of the EchoLink web site found under the **Support and FAQs** link that is on the main page.

Unfortunately it is common for one or more of the UDP tests to fail, which means you may need to make some changes to your Internet router and computer configuration settings to enable UDP port forwarding.

Before getting too wrapped up in this issue, though, try running the **Firewall/Router Test** utility a second time. Newer versions of the EchoLink software (2.0 and above) make use of an alternative UDP data flow mechanism (*port triggering*) that is present in many newer routers. If all the UDP tests pass the second time try continuing with the installation, only coming back to this issue if you later have trouble accepting connections from or connecting to EchoLink nodes.

Firewall / Router Problems

If your Internet connection is being provided via cable or DSL, you most likely have a router somewhere. If you have your own router installed, you can probably get it configured for EchoLink operation. But there are many different brands and models of routers; the method of configuring these routers generally varies by manufacturer and model number. You will need the instructions for your particular router. These typically come with the router, either in print or on a CD. If you don't have the instructions, you may be able to download them from the Internet. You may not have your own router, though, in which case the routing function is probably being provided internal to either the DSL or cable company modem. If that is the case, you may need to contact your service provider for assistance.

You may also have problems with the firewall software that is running in your computer. This could require changes to the

Microsoft firewall settings, or perhaps third party system security software such as Norton or others, which may be installed.

You can easily determine if your computer's firewall is causing a problem by temporarily disabling it and re-running the **Firewall/Router Test** utility. If you do this, be advised that your Internet security is compromised for the period of time the firewall is not enabled. The risk is minimal if you only take a minute or two to run the test and then re-enable the firewall. You can generally find the *Window's Firewall* by looking in the list of programs found under the computer's *Control Panel*.

Configuring Firewalls and Routers

For EchoLink operation, your computer's firewall needs to be configured to accept data directed to TCP port 5200 or to UDP ports 5198 and 5199.

Routers need to be configured to direct data received on UDP ports 5198 and 5199 to any computer, which makes outbound requests over TCP port 5200, or UDP ports 5198 or 5199. Typically there are two methods of configuring routers to do this: UDP *port forwarding* and *port triggering*. If your router provides both methods, you are generally better off using *port triggering* which has the ability to send the data to any computer that made outbound requests on these ports.

Since the instructions for configuring your firewall and router settings are dependent upon your method of Internet connection and type of protection software you are running, it is beyond the scope of this book to cover all possible configurations. For guidance on resolving these issues, refer to the following sources:

- The **Routers and Firewalls** section of the EchoLink web site found at: **http://www.echolink.org/**
- The **Firewall Issues** section of the EchoLink software help files found under the **Troubleshooting** section.
- The **Firewall Solutions** section of the EchoLink web site found under the **Support and FAQs** link on the main page.
- The **http://portforward.com** web site. A link for EchoLink specific topics is found near the bottom of the **Firewall Issues** EchoLink software help instructions.

If you have your router's TCP and UDP ports forwarded correctly, but are still having trouble connecting to EchoLink nodes, it is worth noting that some D-Link routers and perhaps others have a firewall setup item called *NAT Endpoint Filtering*, which can enable or inhibit a port's incoming traffic from being forwarded to the application that opened the port. If your router has this parameter, it should be set to allow port forwarding. In D-Link routers, this is the *Endpoint Independent* selection.

If you are having firewall or router problems, and all this stuff is Greek to you, don't give up! Try and find a fellow ham or an experienced computer user to give you hand. Configuring firewalls and routers can be confusing to the uninitiated.

Windows 7 and Vista Firewall Issues

Both Windows 7 and Windows Vista have a built-in firewall, which is enabled by default. If it's enabled, it can block programs on your computer from being reachable from the Internet. Since EchoLink relies on being accessed via the Internet; it can prevent EchoLink from working unless a special "exception" is added for EchoLink.

The first time you run EchoLink on a Windows 7 or Vista machine, you will most likely encounter a **Windows Security Alert** pop-up window stating, "**Windows Firewall has blocked this program from accepting incoming network connections**". From the same pop-up window you can easily tell Windows to make an exception for EchoLink by selecting the **Unblock** button.

At the following link, the EchoLink web site has additional information related to setting up an exception for EchoLink.
http://www.echolink.org/vista.htm

Call Sign Validation / Authentication

If you have never registered before, when the EchoLink software first runs, the Station List area of the screen will be blank. You will be able to look at the menus and help files, but the software will not allow access to the EchoLink system. You must first have your call sign validated by the system before you are granted full access.

All new users must be validated to ensure that they hold a currently valid amateur license and that they are who they say they are. You certainly would not want someone else using your call sign to gain access to the system. To guard against this, in addition to providing your call sign, you must provide some form of proof of identity, which serves to authenticate who you are. One method is to either fax or upload a scan of your license; several scan image formats are acceptable: TIF, JPEG, PDF, GIF or PNG. If you prefer not to send a copy of your license, there are several alternative methods of authentication to pick from.

For more information on this process and to have your call sign validated and authenticated, go to the main EchoLink web page at **http://www.echolink.org** and access the **Validation** page.

The **Validation** page contains links to the following information:
- **Callsign Validation FAQ**   Answers to frequent questions
- **Authentication**   Explains how to authenticate
- **Validation Documents**   Methods of validation by country

Typically within a day or so after sending the required information, you will receive an email that is used to verify your email address. Click on the link contained in this email to continue with the validation process. Next you should receive another email stating that you have been validated, at that point your call sign has been registered and you have been assigned an EchoLink node number. Your EchoLink software should be fully functional and you can begin making contacts.

Verifying Your Node Has Been Added

Once you have received notification that you have been validated, you can check to see that your call sign / node has been added to the list of stations available on the network. To do this, run the EchoLink software and you should see that the Station List area of the EchoLink window is filled in with available station nodes of all types.

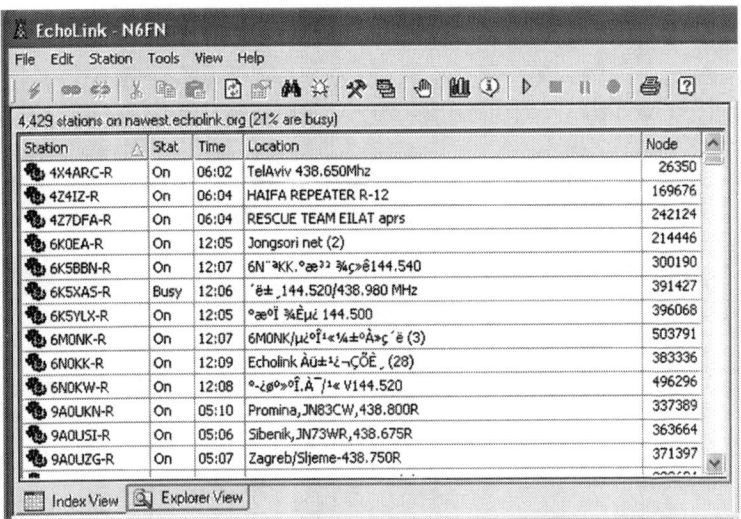

Station List as seen in Index View

Page 28

At the bottom of the station list window, you can select your preferred way of viewing EchoLink nodes. For now select **Explorer View**.

Finding your call sign
Click on the *binocular* icon in the Tool bar and type your call sign in the **Find Station or Location** popup window, as shown in the following figure. Click on **Find** and your call sign / node should come up.

Next hover the mouse cursor over your call sign to view your location, current status (either **On** line, or **Busy**), your local time, and most importantly your new **Node** number.

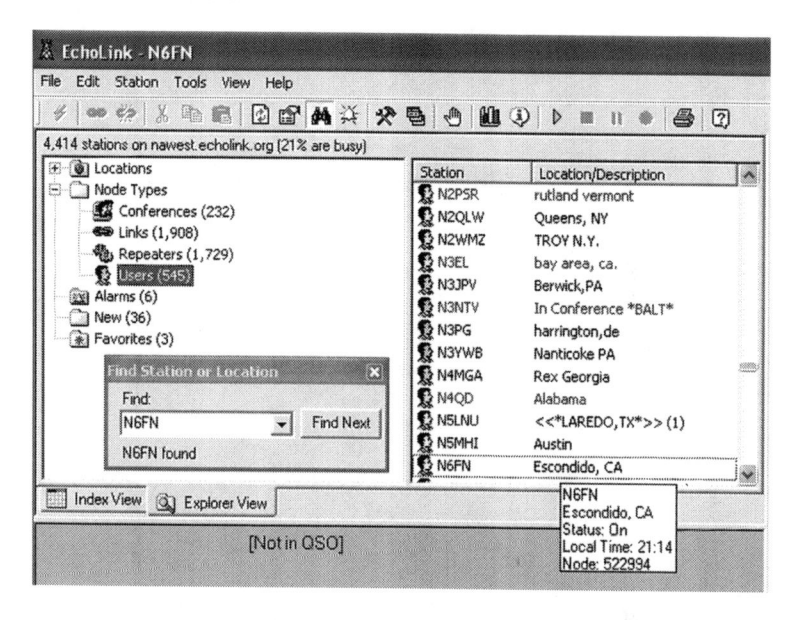

Searching for your call sign, Explorer View

Congratulations, you are now a registered EchoLink user! Your node now appears in the station list of users all around the world. Now you can be contacted by other stations, or initiate your own calls to anywhere in the world. Continue with the next sections, as there are still a few things that need to be configured before actually making contacts.

Setting Audio Parameters and Microphone Type

When running in single-user mode, you should select the type of microphone you are using, either communications or general-purpose. When using a general-purpose microphone, making the selection can improve your transmitted audio by applying a bit of boost to the upper and mid-range frequencies, similar to the frequency response characteristics built into communication microphones.

To select your microphone type, under the **Tools** menu, click on **Setup** and then select the **Audio** tab. Here there are a few other settings that you might want to review as well. Checking the **Auto Sample Rate Compensation** and the **300 Hz TX High-Pass Filter** options might be desirable. To determine if your soundcard sampling rate may need to be adjusted, click on **Calibrate**. Refer to the **Audio** tab's **Help** screen for information on these and the other settings.

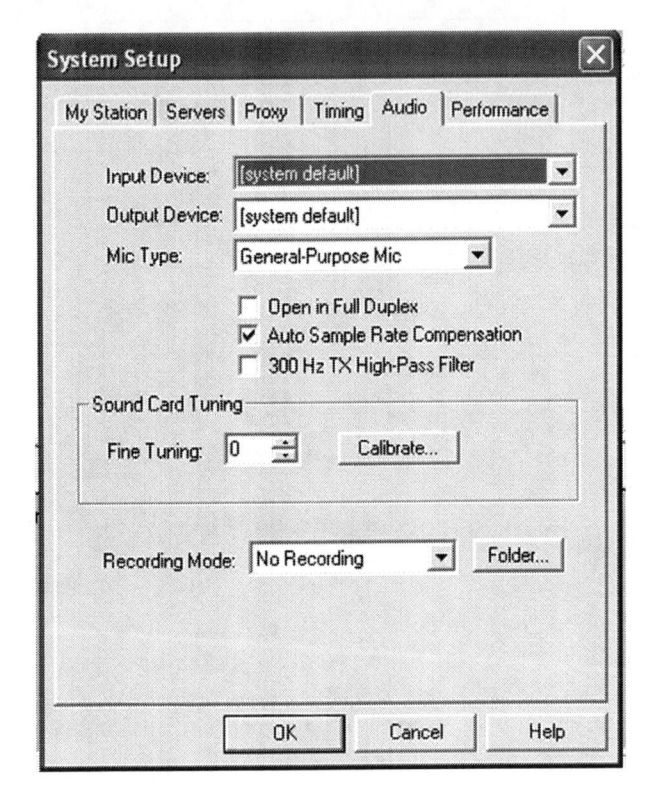

Setting Audio Parameters and Mic Type

Accessing the Test Server and Adjusting Volume Levels

EchoLink has a special Test Server node, useful for verifying that you can receive and send to other nodes and for setting your computer's playback and record volume levels.

To connect to the Test Server and hear a recorded message from Jonathan, K1RFD, under the **Station** menu, click on **Connect to Test Server**. In addition to hearing Jonathan's message, the QSO Status window at the bottom of the screen should indicate that you are connected to the *ECHOTEST* conference server. The server also sent a text message, visible in the upper right of the screen.

To adjust the receive volume, under the **Tools** menu, click on **Adjust Sound Device**, and then select the **Playback** option. This brings up your computer's **Volume Control** window, which depending upon your computer and operating system may appear a bit different than the figure shown below. Both the **Volume Control** and **Wave** sliders are used.

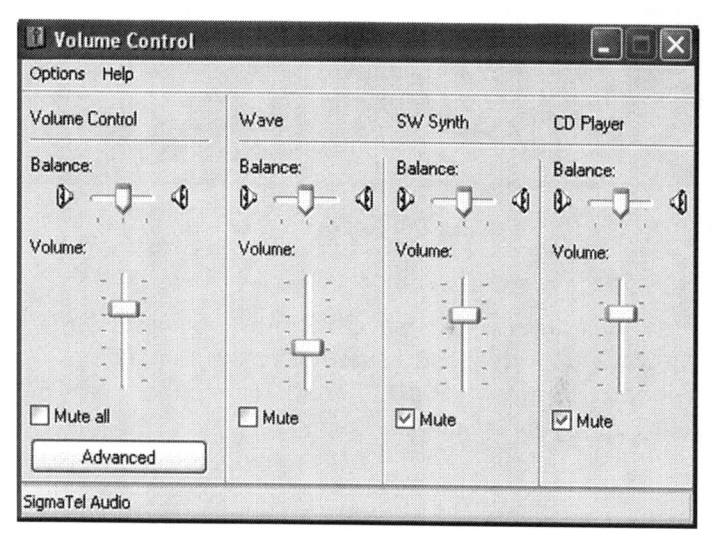

Setting the Received Audio Volume

To replay the Test Server's greeting, first disconnect by clicking on the red disconnect-link button in the Tool bar, and then reconnect. Reconnect and disconnect as many times as you wish to adjust your audio playback level.

To adjust the transmit audio, under the **Tools** menu, click on **Adjust Sound Device,** selecting the **Recording** option. This brings up your computer's **Recording Control** window. If necessary, reconnect to the Test Server. To start recording (actually transmitting to the Internet) click on the "yellow lightning bolt" transmit icon in the Tool bar. This is an alternate action button; press once to transmit, press a second time to terminate the transmission. Adjust your transmit audio so that it is in the upper 2/3 of the blue audio level meter that appears. To prevent audio clipping, avoid reaching full-scale on voice peaks.

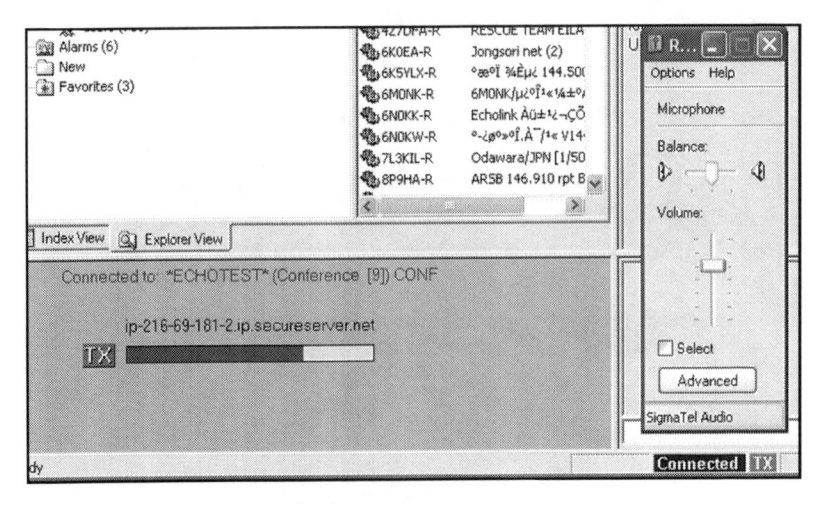

Setting the Recording Level

If the recording level is insufficient when the volume slider is all the way up, click the **Advanced** button under the **Mic** slider and select **Microphone Boost,** **20dB boost**, or similar nomenclature depending upon your operating system.

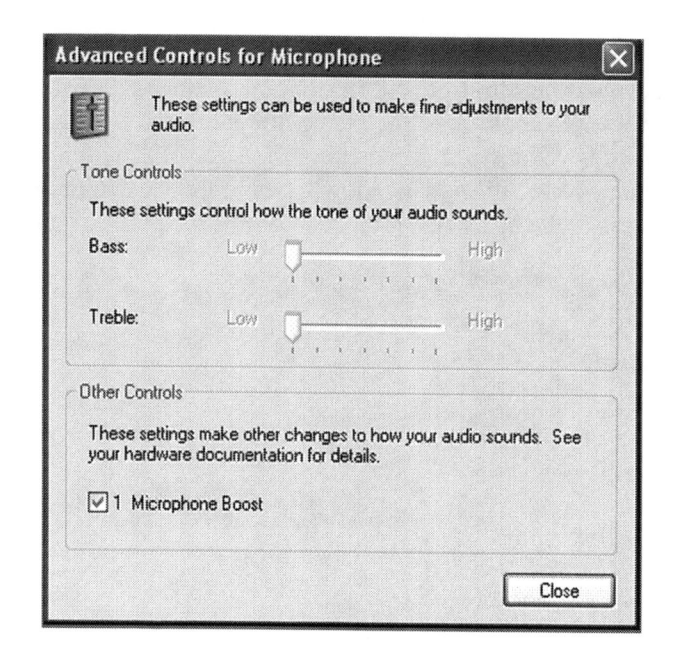

Enabling Microphone Boost

Note: If you are using a headset and can hear a high-frequency hissing noise riding on top of the audio, it's most likely residual noise from your sound card's conversion of the audio. Your headset may have a very high frequency response, great for listening to music but perhaps less than ideal for communication quality audio. Try switching to a different headset, the computer system's speaker or even external speakers. If you really want to use your headset, you may have luck applying some filtering to the received audio, limiting its high frequency bandpass.

Selecting the Transmit Key and its Operating Mode

In addition to the Tool bar's *yellow lightning bolt* transmit icon, you can also initiate transmission with your choice of several different keys on the computer's keyboard. The default is the "Space Bar" but several other keys may also be selected. You can also select if the transmit key operates in a momentary mode (transmit while depressed) or if it acts as a toggle (pressing once to transmit / pressing a second time to stop transmitting). Since EchoLink is half duplex, you must stop transmitting to hear other stations.

To select your transmit key (PTT) options, under the **Tools** menu, click on **Preferences**, and then select the **Connections** tab. In the **Connections** window, click on the **PTT Control...** button to access the **Push-to-Talk Settings** window. Leave the **Keyboard Key** box checked and make a key selection from one of the pull-down options.

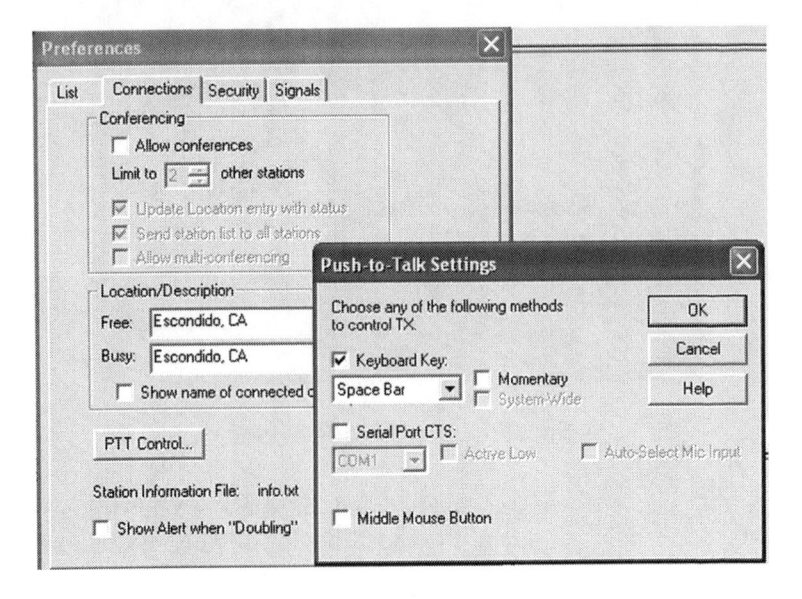

Selecting Transmit Key Options

If you want the key to operate as a momentary switch, much like a microphone's PTT button, check the **Momentary** box. Otherwise leave it unchecked to select toggle mode operation: one press to enable transmit, a second press to quit transmitting.

You can verify how this is working by using our old friend the **Test Server**, found under the Tool bar's **Station** menu. After each transmit test be sure to allow time for the Test Server to play back the audio it recorded – which is probably nothing, but it still needs to play it back.

One thing to remember, *in the default mode* the TX key will only function when the EchoLink window is currently selected. If you want to use other Windows programs while operating EchoLink, you can check the **System-Wide** box. But the System-Wide function is not available when the space-bar key is selected, only when one of the other keys is selected. If you chose to operate with one of the other key selections, that key will lose its usual function. So pick a key you would not normally be using while operating on EchoLink.

Customizing Your Station Information

Whenever you establish a connection to another station, EchoLink transmits a text message containing your station information to the other station's received text window, located in the upper right section of the main EchoLink window. As an example, the following figure shows the text message that the Test Server sent when you connected. In addition to the station's identification (shown as "CONF Audio test server" in the top line in this example) nodes may provide a welcome message or other information when you connect.

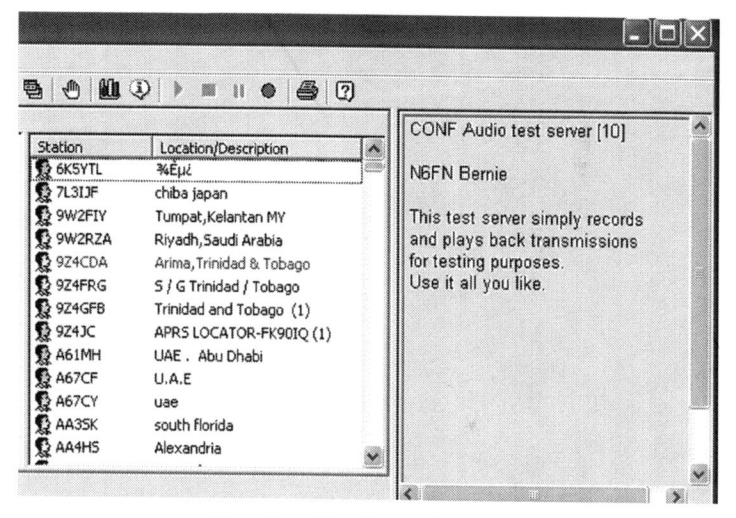

Viewing Station Information

To examine the default text that EchoLink generated for your station when you registered, under the **Tools** menu, click on **Preferences**, and then select the **Connections** tab. At the bottom of the **Connections** window, click on the **Edit** button to access the **Station Info** edit window. Here you can customize your station's information, adding a personal comment or other information as desired.

Verifying How Your Name Appears

You can verify, and edit if you like, how your name will be displayed at other stations. Under the **Tools** menu, click on **Setup**, and then select the **My Station** tab. Here you can edit your name so that it will be displayed whichever way you like.

Chapter 5: Operating EchoLink

A great thing about EchoLink is the flexibility of being able to make contacts using either your radio or computer. You probably don't need new equipment and most likely can use what you already have.

Think about it. Most of us operate our VHF/UHF capable radios through repeaters, either around town or when traveling cross-country. Routinely making contacts while walking around with an HT, while mobile in our vehicles, or from a home base station. Providing we are within range of either an EchoLink equipped simplex link or repeater, we have an opportunity to use those same DTMF equipped radios to make contacts world wide, not just to those stations within a local repeater's coverage area.

That's pretty cool, but here is the icing on the cake! What happens when we are not within range of a repeater? Without EchoLink, that would be it – no reliable way of making VHF/UHF contacts. Yet, by using a computer (or an iPod, iPad, iPhone) equipped with EchoLink, providing we have Internet access, either directly connected, or through a publicly provided wireless hotspot, we can still make contacts all over the world.

Initiating a QSO Via Computer

Once the EchoLink software has been installed and set up, and you have properly adjusted the received and transmitted audio levels using the Test Server, you are cleared for takeoff!

There are several ways of finding a station, repeater or conference server to make contact with:

- Searching manually through the Station List in the EchoLink window, viewable either by Location or Node Type.

- Searching the Station List automatically by using the tool bar's *binocular* find icon to search by country, state, city, etc.

- Searching by location using the *EchoLink Directory* found at **http://www.brenet.com/Echolink.htm** Searchable by station / node type, state and frequency band.

- Searching by lat / lon, grid square, city, state or country using the *EchoLink Node Status* directory located at **http://www.echolink.org/links.jsp** Results sorted by distance.

- Searching through *EchoLink Nets Yahoo Group* postings found at **http://groups.yahoo.com/group/Echolink_Nets/**

Once you have found a station via the Station List, it can be connected to as described below. If you found a station's call sign using a web page or some other reference, you can either search for it in the Station List using the tool bar's *binocular* find icon, or enter it directly by selecting the **Connect to...** option found under **Station** in the menu bar.

Connecting via the Station List

The Station List shows stations that are currently logged on. Selecting the **Explorer View** tab displays folders of stations sorted by geographical **Location** and by **Node Types**. The **New** folder contains stations that have logged on within the last five minutes.

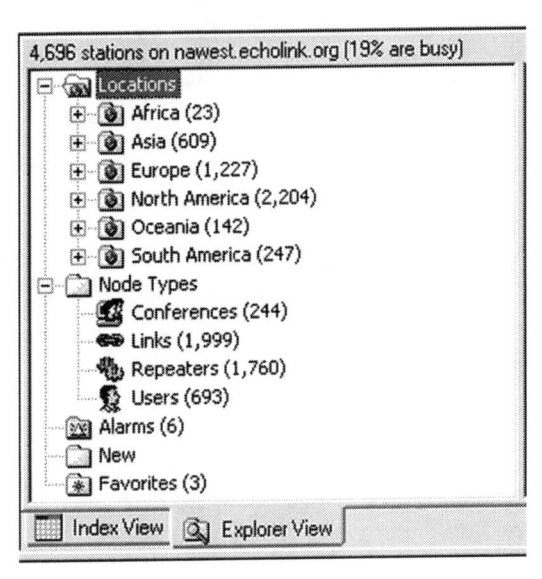

```
4,696 stations on nawest.echolink.org (19% are busy)
└─ Locations
    ├─ Africa (23)
    ├─ Asia (609)
    ├─ Europe (1,227)
    ├─ North America (2,204)
    ├─ Oceania (142)
    └─ South America (247)
└─ Node Types
    ├─ Conferences (244)
    ├─ Links (1,999)
    ├─ Repeaters (1,760)
    └─ Users (693)
└─ Alarms (6)
└─ New
└─ Favorites (3)

    Index View    Explorer View
```

You can select individual stations by opening a desired folder. Stations showing up in blue text are already in a QSO, "Busy" and cannot be contacted. Hovering your mouse over a station displays its status, local time and node number.

Connecting to a station showing in the list could not be easier; select a desired station and double-clicking on it. Alternatively, you can select a station and then click on the tool bar's *green connect link* icon.

If you attempt to connect to a station / node that is already busy, the QSO status window will display a "**Cannot connect to XXXXX – Busy**" message. If your connection is accepted, you will see a "**Connected to: XXXXX**" message, and additional station information may appear in the upper right received text box.

If you did not get connected and a "**Timed Out**" message was displayed after about 30 seconds, it could be for one of several reasons:
- The station may have just gone off line in the last few moments,
- The station may have just established contact with someone else,
- Your firewall might be preventing you from receiving data from that station. Refer to the earlier section on firewall issues.

When the other station is transmitting, a green **RX** icon is displayed in the bottom right status bar. You can transmit only when the other station is not transmitting. While you are transmitting a red **TX** icon is displayed in the status bar.

Connected and Receiving

Connecting to Conference Servers

Conference servers allow linking multiple stations together, each station becoming a participant in a multi-party QSO.

When viewing conference servers and simplex link conference servers in the Station List, the numbers enclosed in brackets following the station's description indicate how many users are currently connected and the maximum number of users that the server is configured to accept.

Station	Location/Description
1-NIPPON	3562Kz SSB ToneSQ-L [8/59]
10WATTS	10WATTS GROUP THAILAND [16
145MHZ	www.145mhz.org [12]
1AREA-JP	JAPAN [1/88]
29MHZ-FM	29MHz FM Conference [4/30]
902_HUB	Private conference [8/15]
ACORES	S. Roque do Pico [4/50]
AGUALINA	In Conference *LK2*
ALARA	Australian Ladies C [0/500]
ALASKA	Alaska [3/500]
ALERTERC	Private conference [2/100]
ALLJAPAN	* ALL-JAPAN Net * [8/500]
ALTOHOKU	ALL Tohoku-Net [16/59]

Conference Servers Station List

Operating Procedures

Following proper operating procedures will enhance your operating experience and avoid irritating others. These procedures are equally applicable whether connecting via your computer or your transceiver. Once a connection has been established, it's best to adhere to the following procedures:

- When first connecting to a simplex link, repeater or conference server, listen for a minute before transmitting, as a QSO might already be in progress.

- Using your call sign, announce you are there and state your intended purpose. Perhaps asking if anyone is available to verify you are getting through, or simply "This is XXXXX listening".

- It is considered rude to connect to a repeater or simplex link without announcing your presence. Unless the node's sysop has turned it off, an over-the-air voice announcement that says

"**Connecting to EchoLink** (your call sign) **Connected**" has announced your presence. You may not hear this announcement on your end. When you disconnect, a similar disconnect announcement is made. Everyone within range, or linked to the repeater, will hear these announcements. If you connect – be courteous and announce your presence and intention.

- For the same reasons, it also follows that you don't want to repeatedly connect and disconnect to the same node. You may bombard the node with connect / disconnect announcements.

- When someone comes back to you, pause several seconds before answering. Because of the extra delays through linking transceivers and repeater timer delays, pausing between transmissions is especially important. PAUSE, PAUSE, PAUSE

- When starting transmission, either by pressing a button or keying a mic, wait a second or two before talking. This allows time for receivers linked to repeaters to stop transmitting and start receiving. PAUSE, PAUSE, PAUSE

- Keep your transmissions short. Repeaters and simplex links frequently have maximum time outs configured, generally three minutes and perhaps less. (Note: Unless the sysop has implemented external TX timeout hardware, if you exceed EchoLink's maximum transmit time, you will be automatically disconnected from the node.)

- Do not tie-up heavily used conference servers with extended one-on-one conversations, instead set up a direct connection.

- If you start a conversation with someone not local to the remote node, (neither of you are local to the repeater) set up a direct connection to prevent tying up the repeater with non-local traffic.

- If you are operating on a repeater and have linked to another repeater, don't leave a repeater-to-repeater link open for extended periods of time. The exception is when the operators of both repeaters intend to create a semi-permanent or permanent link between the two repeaters.

- When the QSO is finished, disconnect from the remote station either by clicking the red *broken link* icon when operating from a computer, or by sending the DTMF disconnect command (usually the **#** key, sometimes **73**) when operating over an RF link.

Direct Computer-to-Computer Contacts

In addition to contacting stations via repeaters and conference servers, you can also make direct computer-to-computer contacts without going over an RF link at either end of the QSO. While technically this is not governed by FCC rules, it's typical for both stations to adhere to the customary ham radio operating procedures: usually giving call signs and indicating when you are turning it over to the other station.

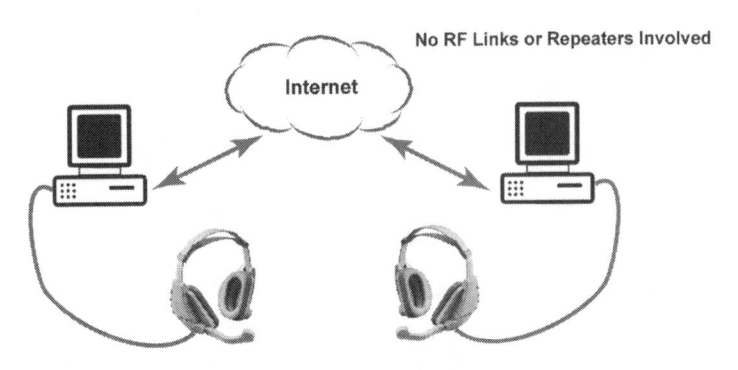

PC-to-PC VoIP Communication

While these contacts may not be "ham radio," they can be convenient for keeping in touch or having long conversations. There is also a measure of privacy in computer-to-computer links that is not present when operating on conference servers or repeaters.

Providing the station you wish to contact is on-line, and you know the call sign, initiating contacts is performed as described in the above **Connecting via the Station List** procedures.

As a side note, if you like the idea of being able to make voice contacts over the Internet, you can also use a free Internet service called *Skype* to make contact with non-hams. *Skype* even allows you to use your computer to make calls to phones, set up conference calls of three or more, and optionally transmit live video. *Skype* can also be used for *Instant Messaging*, *File Transfers* and *Video Conferencing*. While computer-to-computer contacts are free, there is a per-minute fee for calling either landline or mobile phones. It's not ham radio, but can be useful for economically keeping in contact with family members and friends.

Full information can be found at:
http://www.skype.com/intl/en-us/home

Chat / Texting

When you and the other station are operating computer-to-computer, you can also send typed messages back-and-forth. Once the connection has been established, a short message typed in the small text message box to the left of the **Send** button can be sent by pressing **Send**. Note that you can type the message while listening in **RX** mode, but it is not transmitted until you press the **Send** button.

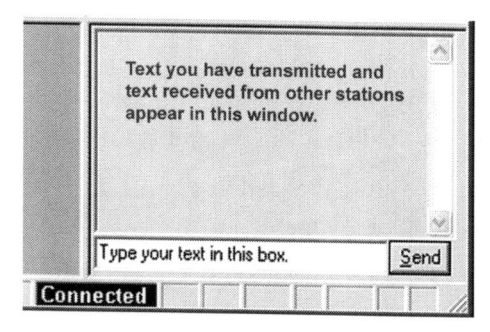

Entering and Viewing Transmitted Text

Your own message and any reply messages from the distant station will appear in the Chat window directly above the text message box used to format your messages.

When operating on a conference server, transmitted text messages are sent to all single user participants (those with computers).

Disconnecting

After completing a QSO, or if you were unable to raise someone, it's important to terminate your connection to the remote station / node.

When using the EchoLink computer program, click the *red broken link* icon on the tool bar to disconnect. You should hear a disconnect beep signal, and the **[Not in QSO]** status message should appear near the bottom of the screen.

If operating via DTMF commands using a transceiver, send the DTMF disconnect command (usually the **#** key, and sometimes **73**).

Keeping Track of Favorites

The **Favorites** folder in Explorer View is for saving the call signs of stations you may wish to contact on a regular basis.

There are several ways of adding stations to this folder:
- From the **Station** List, right click on a selected station's call sign and select **Add to Favorites** from the pop-up window.

- From the **Station** List, right click on a selected station and select **Copy** from the pop-up window and then open **Favorites** folder and paste it into the list of stations in the folder.

- Using the mouse, drag the station's call sign from the right hand station list, dropping it into the **Favorites** folder.

- To enter the call sign of a station that is not currently logged in to EchoLink (not in the Station list), open the **Favorites** folder, right click in the **Station** list window and select **New Callsign** from the pop-up window. Then type in the call sign. The station will then appear in the list showing a status of **off**.

If your list of favorite stations gets to be a bit long you can organize your favorite call signs into sub-folders.

Calls can be deleted from your **Favorites** folder by right clicking on the call sign and selecting **Delete** from the pop-up window.

Setting Station Alarms

In addition to keeping track of stations in the **Favorites** folder, by adding stations to the **Alarm** folder you can be alerted whenever specific stations come on-line. An alarm will sound and a pop-up window appears whenever one of the stations in the **Alarm** folder either logs on, logs off or changes status.

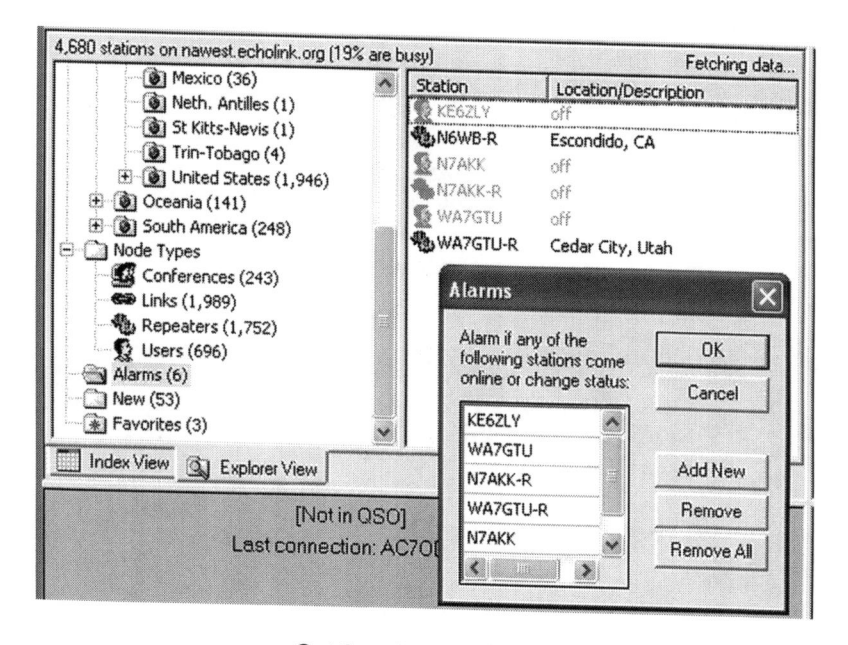

Setting Station Alarms

Station call signs are added to the **Alarm** folder using the same techniques described above for adding stations to the **Favorites** folder. In addition, you can use the **Alarms** selection found under the **Tools** menu to manually enter (or delete) call signs in the **Alarm** folder.

Low-Bandwidth Indicator

If the available Internet data rate is insufficient to keep up with the amount of voice data being sent by EchoLink, a yellow triangle with an exclamation mark in the middle is displayed at the bottom right of the screen.

Even when on a relatively fast Internet connection, if you are hosting multiple users in conference mode, the insufficient bandwidth warning might come up. The solution is to either get a faster connection or limit the number of users. This error might also occur in single-user situations if you are on a dial-up line.

Call Sign Log

The history of stations you have contacted is saved in the **Callsign Log**. In addition to recording call signs, it also records the time you connected and disconnected from the node.

The log can be viewed by using the **Callsign Log** selection found under the tool bar's **View** menu. While the list of calls is being displayed, you can use the **Find** button to search for specific stations or other information.

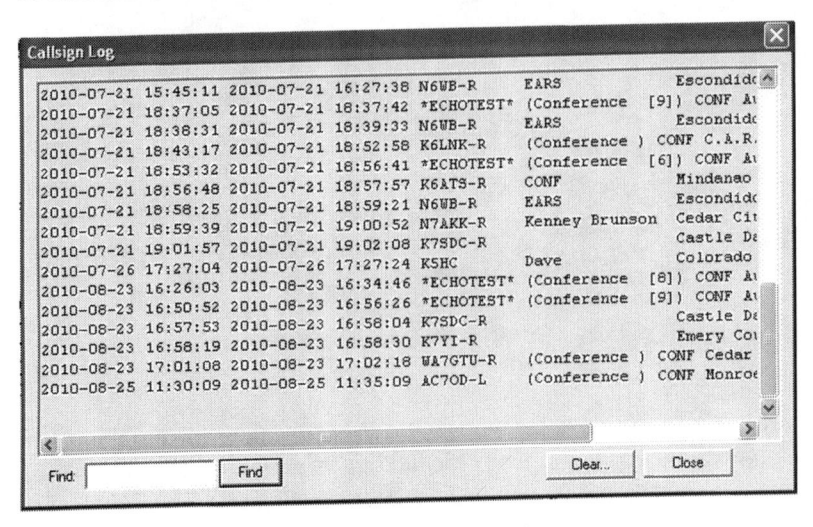

```
Callsign Log
2010-07-21 15:45:11  2010-07-21 16:27:38  N6WB-R      EARS                    Escondido
2010-07-21 18:37:05  2010-07-21 18:37:42  *ECHOTEST*  (Conference )  [9]) CONF A\
2010-07-21 18:38:31  2010-07-21 18:39:33  N6WB-R      EARS                    Escondido
2010-07-21 18:43:17  2010-07-21 18:52:58  K6LNK-R     (Conference )  CONF C.A.R.
2010-07-21 18:53:32  2010-07-21 18:56:41  *ECHOTEST*  (Conference )  [6]) CONF A\
2010-07-21 18:56:48  2010-07-21 18:57:57  K6ATS-R     CONF                    Mindanao
2010-07-21 18:58:25  2010-07-21 18:59:21  N6WB-R      EARS                    Escondido
2010-07-21 18:59:39  2010-07-21 19:00:52  N7AKK-R     Kenney Brunson          Cedar Cit
2010-07-21 19:01:57  2010-07-21 19:02:08  K7SDC-R                             Castle De
2010-07-26 17:27:04  2010-07-26 17:27:24  K5HC        Dave                    Colorado
2010-08-23 16:26:03  2010-08-23 16:34:46  *ECHOTEST*  (Conference )  [8]) CONF A\
2010-08-23 16:50:52  2010-08-23 16:56:26  *ECHOTEST*  (Conference )  [9]) CONF A\
2010-08-23 16:57:53  2010-08-23 16:58:04  K7SDC-R                             Castle De
2010-08-23 16:58:19  2010-08-23 16:58:30  K7YI-R                              Emery Col
2010-08-23 17:01:08  2010-08-23 17:02:18  WA7GTU-R    (Conference )  CONF Cedar
2010-08-25 11:30:09  2010-08-25 11:35:09  AC7OD-L     (Conference )  CONF Monroe

Find:            Find                              Clear...    Close
```

Viewing the Callsign Log

Also under the **View** menu, you can select **Callsign Log** then **Edit**, which opens up a Windows **Notepad** text editing window allowing you to view and edit the data as desired. The edited text file can also be saved to disk, perhaps useful for keeping a permanent record of contacts made.

DTMF Keypad Commands

Providing you are within range of either an RF simplex link or RF repeater link, you can use your DTMF equipped VHF/UHF transceiver to connect to: individual users, repeaters and conference servers. Using your radio's DTMF keypad to enter commands, you can check the status, bring-up and then close connections to distant EchoLink nodes.

Sysops that control RF simplex and repeater Links can either accept EchoLink's default DTMF commands, or create their own DTMF commands. Custom commands might be created as part of an effort to keep unauthorized users from accessing the system, for preventing interference with other DTMF functions, or for other reasons. If the RF simplex or repeater Link that you are trying to use does not respond to EchoLink's default commands, you will need to contact the node's sysop or visit the sysop's (or sponsoring club's) web page to find out the correct procedures to use.

The following table shows the most frequently used DTMF command codes. These are EchoLink's default codes; node operators have the option to modify them as desired. Besides the codes shown in the table, EchoLink supports an additional seven codes that make randomly selected connections to different types of nodes. If you wish to make use of the random connect commands, refer to the **DTMF Functions** section of the **EchoLink Help** system.

Two of the default DTMF commands (*Connect-by-Call* and *Query-by-Call*) allow for the entry of a station's call sign. Since the DTMF keypad does not directly support entry of alpha characters, a two-press per character keying system has been developed for keying alphabetical call signs. Refer to Appendix B for details on how to enter alpha characters using a DTMF keypad.

Default DTMF Command Codes

DTMF Code	Command	Description
*	Play ID	Plays brief identification message
#	Disconnect	Terminates current connection
0511	Listen-Only On	Prevents local RF transmissions from going over the Internet to other nodes.
0510	Listen-Only Off	Terminates Listen-Only Operation
06+num	Query by Node Number	Looks up a station by Node Number and reports back its call sign and status.
07+call+#	Query by Call Sign	Looks up a station by Call Sign and reports back its node number and status.
08	Status	Determines if node is already connected. Reports call signs of connected stations.
09	Reconnect	Reconnects to the station that was most recently disconnected.
Number	Connect by Node Number	Connects to an EchoLink station by specifying the node number, either 4,5 or 6 digits.
C+call+#	Connect by Call Sign	Connects to an Internet station by specifying its alphanumeric call sign.
9999	Test Server	Connects to the Echo Test Server

Note: For terminating a connection, some Sysops have selected to use **73** instead of EchoLink's default **#** symbol.

Beyond the Default Codes shown above, node operators have the option of creating additional DTMF commands that act as shortcuts to specific stations. You must contact the node's sponsoring Sysop or Club to find out what these are. Frequently they are published on the sponsor's web page.

Initiating a QSO with a Radio

To initiate a QSO via radio, you first need to locate an RF simplex or repeater Link to use as a gateway for accessing the EchoLink network. Links can be found by checking with your local ham radio store, club or friends in the area. In addition, you can also search the Internet for RF links local to your area. The two Internet sources previously mentioned for making computer contacts are also useful for determining the RF link's frequency offset and access tone information.

- The searchable by location *EchoLink Directory* found at **http://www.brenet.com/Echolink.htm** Searchable by station / node type, state and frequency band.

- The searchable by lat / lon, grid square, city, state or country *EchoLink Node Status* directory located at **http://www.echolink.org/links.jsp** Results sorted by distance.

Once the RF link's frequency, offset (if any) and access tone have been determined, for future convenience they should be programmed into your transceiver's memory as usual. This frequency will be used as your gateway for gaining access to the EchoLink network.

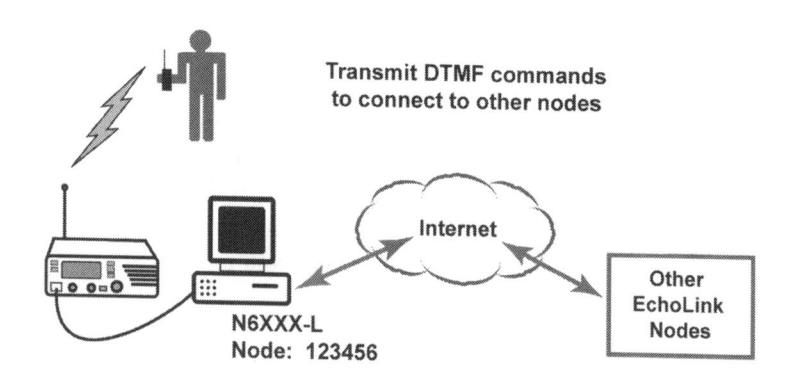

Simplex Link Gateway Operation

Secondly you will need the call signs and node numbers for any individual users, repeaters or conference servers that you wish to connect to. These can be looked up using the two web pages mentioned above, or by searching the station list in the EchoLink software. When using a DTMF keypad, though, it's simpler to use the station's node number rather than its call sign. Since you will presumably be using your transceiver when you are without Internet access, information for stations to be contacted should be determined and recorded in advance.

If you don't remember how to activate DTMF tones on your transceiver, you may need to consult your radio's user manual, or its Nifty Guide. Unless your transceiver's DTMF capability has been turned off, or set to autodial from DTMF memories, the typical procedure for sending DTMF tones is to hold PTT down while pressing the desired DTMF keys.

Once you have your radio set up and know the node number of a station you wish to call, the following example outlines steps to be followed to initiate a contact.

Example – Establishing Connections to Other Stations

While this example establishes a connection to the Test Server, the procedure will work for any station you wish to contact. In step 4, use the node number of the station you wish to contact, instead of 9999 for the Test Server.

1. First listen to make sure the local RF simplex link or repeater frequency you intend to use as a gateway is not already in use.

2. To verify that EchoLink is running on the Link you are using as a gateway, using your DTMF keypad, send the **08** command. If EchoLink is up and running, the response should be "**Not Connected**" or "**Connected to XXXXX**" where XXXXX is the call sign of the connected station.

3. If a station is connected, announce your presence and see if you get a response. If not connected, announce your intention to access EchoLink, saying something like: "N6XXX accessing EchoLink". Wait a moment to see if any stations respond before proceeding with entering DTMF commands.

4. To establish a connection, enter the Test Server's node number: **9 9 9 9** . After a short delay, you should hear EchoLink responding with "**Connecting to Conference E C H O T E S T**", followed a little later by "**Connected**" if you were successful. If the specified node is not currently on-line, you will hear "**Not Found**". Wait until you hear the **Connected** response before transmitting your call.

If you hear a **Disconnected** right after trying to connect, the node may have a firewall or Internet problem. You will have to try a different node.

5. After a few seconds you should hear the Test Server's greeting: "**Welcome to the EchoLink Test Server ...**" For other nodes, you may or may not get a greeting.

6. In the case of the Test Server, if you make a voice transmission, after a short delay it should be echoed back to you. For other contacts, first listen to make sure that there is not a QSO in progress. When clear, you can make your call on the connected node. Remember to pause an extra amount of time between transmissions and to ID as you usually would when working through a repeater.

7. To terminate the connection, press **#**. You should then hear EchoLink respond with "**Conference E C H O T E S T Disconnected**". (Many nodes use **73** instead of **#**.)

Important: It's your responsibility to terminate any connections that you might establish. When operating mobile and approaching the fringe area of the node you are using to access EchoLink, make sure you terminate the connection before you are out of range. (As a side note, if necessary anyone can terminate a connection by transmitting the DTMF **#** code, or if applicable **73**.)

Initiating a QSO via a Wireless Hot Spot

If you are away from home and are unable to access either an RF simplex link or EchoLink equipped repeater with your radio, you may still be able to access the EchoLink network with your laptop via a *wireless community network* or *hotspot*, as commonly offered in hotels, Internet cafés and other public areas.

A problem with using public hotspots is that they generally won't allow *peer-to-peer* Internet connections. They allow you to access web pages and email as usual, but attempts by other Internet nodes to connect to your laptop will most likely be blocked. Since EchoLink nodes normally operate by connecting to each other as peers, sending text and voice signals directly to each other, this would seem to preclude EchoLink operation. There is a solution, nevertheless, via the use of EchoLink proxy servers.

EchoLink proxy servers are Internet connected computer systems that host a special proxy server version of the EchoLink software. By enabling proxy server operation on your laptop, you will be able to use EchoLink from hotspots without requiring a peer-to-peer connection. This allows your laptop to establish a normal outbound connection to the proxy server, which in turn will establish the necessary peer-to-peer connections to other EchoLink nodes. In effect, the EchoLink software is split into two pieces, one that runs on your laptop, and another that runs on the remote proxy server.

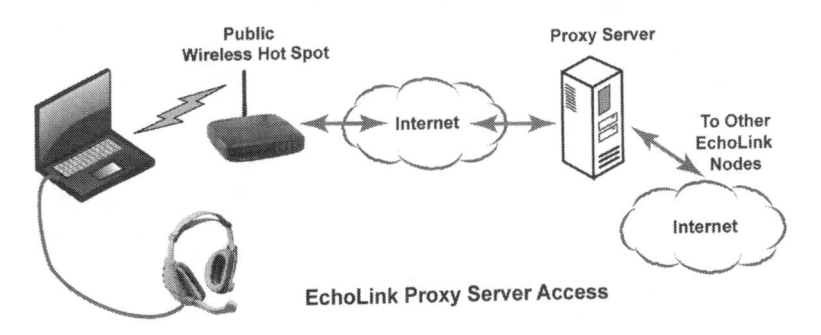

EchoLink Proxy Server Access

There are some drawbacks to using an EchoLink proxy and you should not use it as your normal EchoLink access method:
- There are a limited number of proxy servers
- Proxy Servers can only host one client connection at a time
- Servers are best reserved for those using public hotspots
- Extra network delays are incurred going through the server
- A greater susceptibility to Internet network congestion

You must be in **Single User** mode, not **Sysop** to establish a proxy server connection. To enable proxy server access, on the tool bar select **Tools**, select **Setup** and then select the **Proxy** tab.

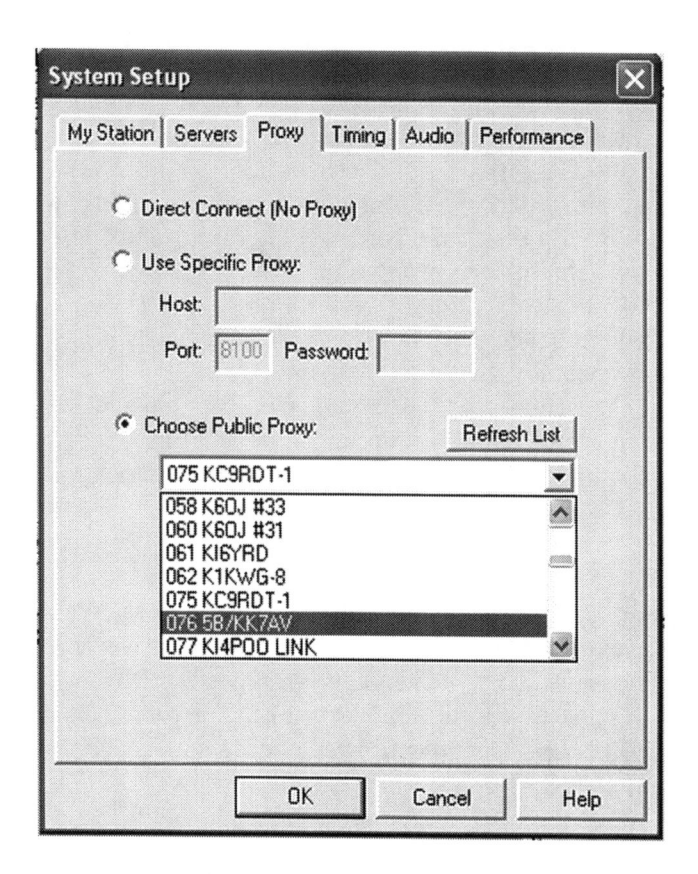

Selecting a Proxy Server

For normal EchoLink operation, **Direct Connection (No Proxy)** should be selected. After you are done making contacts using proxy server, be sure to restore the **(No Proxy)** setting.

Unless you are going to set up your own private proxy server, select the **Choose Public Proxy** setting. Select an available server from the pull down list. To minimize Internet delays, it's probably best to pick a server close to your geographical area. For convenience, the list is automatically sorted in order of increasing distance. Be patient, as it may take a minute or so to establish the connection and update the station list.

Once the connection has been established, and the station list has been displayed, you can make connections to other nodes and operate as usual.

Only servers currently on-line and not busy are shown in the list. If it's been awhile since you selected the **Proxy** tab, click on the **Refresh List** button to update the server list.

Additional proxy server information is available in the EchoLink application's **Help** files and on the EchoLink web page: http://www.echolink.org/proxy.htm

A list of currently active public proxy servers is available at http://www.echolink.org/proxylist.jsp
While not essential for operation, this list provides additional information regarding port numbers, host addresses, current status and comments from the administrators of the proxy server.

EchoLink for iPod Touch, iPad and iPhone

A free iPod Touch, iPad and iPhone EchoLink application allows you to access the EchoLink system from nearly anyplace where WiFi networking is available. If you have an iPhone or 3G iPad you can also use it to access EchoLink over the cellular data (3G or EDGE) network. The *EchoLink for iPhone* application is available from Apple's App Store.

The *EchoLink for iPhone* application provides basic EchoLink features, such as managing the station list, connecting and disconnecting, and transmitting and receiving voice and text. Unlike the Windows version, conferencing cannot be enabled and there is no Sysop mode.

For additional information, operational instructions and frequently asked questions regarding the *EchoLink for iPhone* application refer to http://www.echolink.org/faq_iphone.htm

EchoLink for Android Devices

A free Android compatible application is available allowing you to access the EchoLink system from nearly anywhere where WiFi or 3G networking is available. EchoLink for Android is available free of charge from the *Android Market*; tap the *Market* icon on your device and search for EchoLink. The app is also available from other Android repositories, for devices that do not have access to the *Android Market*.

The Android application focuses on the basic features of EchoLink, such as managing the station list, connecting and disconnecting, and transmitting and receiving voice and text. Unlike the Windows version, conferencing cannot be enabled and there is no Sysop mode.

For additional information, operational instructions and frequently asked questions regarding the Android compatible application refer to
http://www.echolink.org/faq_android.htm

EchoLink Nets

There is an *EchoLink_Nets* Yahoo Group that keeps track of EchoLink nets and special events. This is a good one for making contacts and keeping up with EchoLink related activity.
http://groups.yahoo.com/group/Echolink_Nets/

Useful Web Pages

As you operate you might find the following web pages helpful:

EchoLink Node Status
An EchoLink node locator page where you can search for currently active nodes by lat / lon, grid square or by city and state. Useful for locating stations while traveling, or for making contacts to specific parts of the country.
On **http://www.echolink.org/** click the **Link Status** button.

Nodes Currently Logged On
A list of currently logged in EchoLink nodes, searchable by conferences, links and users can be found on the main EchoLink web page.
On **http://www.echolink.org/** click the **Current Logins** button.

EchoLink Directory
An EchoLink node locator page where you can search by node type (conference, repeater or link), state or province, and by frequency band. Lists all nodes satisfying search criteria by state or province. Useful for locating stations while traveling, or for making contacts to specific parts of the country.
http://www.brenet.com/Echolink.htm

EchoLink **Yahoo Group**
Discussions of EchoLink in general, connection problems, simplex links, repeaters, link interfacing and more.
http://groups.yahoo.com/group/echolink/

Chapter 6: Establishing an RF Node

Establishing your own RF simplex or repeater node is a bit more involved than setting up as a single-user, requiring the following:
- Selecting Sysop mode and configuring the Sysop settings
- Validating –L or –R variants of your call sign
- Acquiring a PC-to-transceiver hardware interface
- Dedicating a computer to operating the node
- Dedicating a suitable transceiver to the node's operation
- Configuring software for the transceiver to PC interface
- Configuring the transceiver's audio and transmit parameters
- Selecting a method for keying transceiver transmission
- Preventing repeater IDs & squelch tails from going to the Internet
- Preventing Windows system sounds from going to the Internet
- Locating a non-interfering clear frequency for a Simplex link
- Seeking permission from the repeater's administrator

Selecting Sysop Mode

If you wish to establish your own EchoLink RF simplex node or an RF link to an existing repeater, you need to enable **Sysop** mode. Assuming you have already been operating as a single-user, to change to sysop mode, access the **Tools** menu, click on **Setup**, and then after selecting the **My Station** tab, set the mode to **Sysop**.

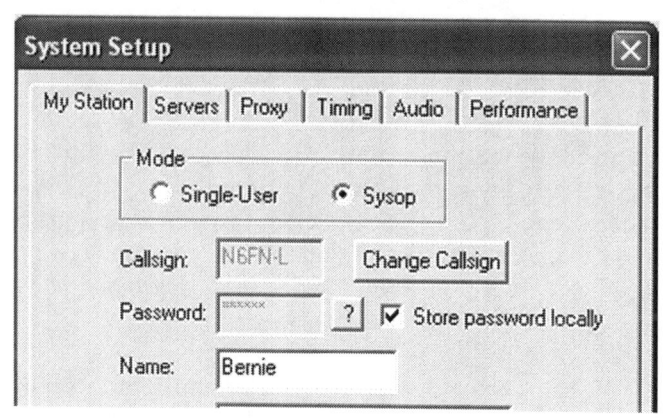

Enabling Sysop Mode

If you have not already done so, you will need to add either a **–L** or **–R** variant of your call sign to the EchoLink database. Click on the **Change Callsign** button to make the change. For establishing an RF simplex link, create a **–L** version of your call sign. For setting up an RF Link to a repeater, use **–R**.

You will not lose your base call sign by creating a call sign variant; this process just adds a new one to the database. When you again desire to operate as a single-user, come back to this screen, reselect **Single User** and enter your call sign. When making these changes you should be on-line to the Internet so that your changes can be tracked with updates to the EchoLink database.

Once you have entered the new call sign as described above, you will need to have it validated, much like when you originally validated your base call sign. To validate your new call sign, access EchoLink's validation page at **http://www.echolink.org/validation/** . On this page, enter your new call sign in the box provided and click on **Continue** to complete the process on the next page that is displayed. After a short while you will receive an email. By clicking on the link provided in the email you will be given your choice of methods to use for validation. Using your previously established password is the simplest method. A few moments after entering your password, an Internet web page should be displayed, indicating you have been validated.

Interfacing a PC to a Transceiver

To set up your own Simplex link or to add EchoLink capability to an existing repeater, you will need to interface a transceiver to a computer system. When selecting a PC it would be well to consider a laptop since its battery can provide some protection against brief power outages. A laptop also provides some space and power usage savings.

Before you can continue with configuring the **Sysop** settings, you need to determine the type of PC-to-transceiver interface you will be using, as it affects the selections you will be making.

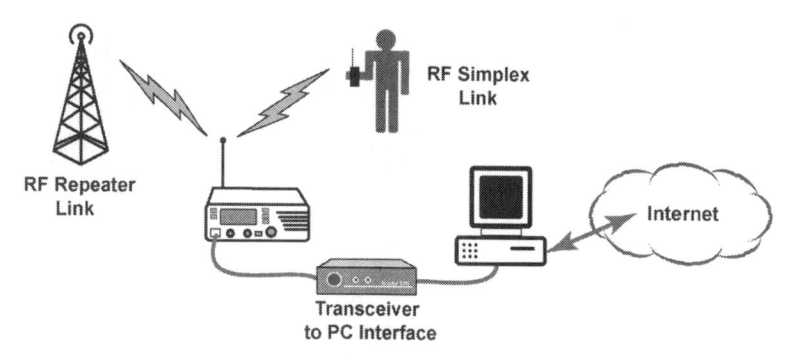

RF Link - Transceiver to PC Interface

As shown above, to create either an RF simplex or RF repeater link, some kind of hardware interface and cabling is required between the link transceiver and the PC running EchoLink. The same interface hardware can generally be used for creating either kind of link.

In concept the required interface is quite simple. Since the computer's sound card provides the audio signal that modulates the transceiver, it follows that the sound card's Speaker Out jack (or preferably if your computer has one, the Line Out jack) is connected to the MIC input of the transceiver. Likewise the transceiver's Speaker Output is connected to the sound card's Microphone Input (or Line In if it has one) so that received signals can be transported over the Internet.

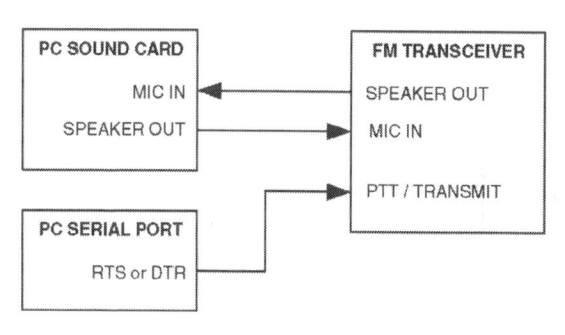

Simplified PC to Transceiver Interconnect Diagram

As illustrated in the above diagram, the basic interconnect concept is quite straightforward. But because of signal level differences, potential ground loop problems, sound card volume setting issues, differences in radios used for the link transceiver, unwanted noises being transported over the EchoLink network and other issues, getting things connected and working properly can become more complex in actual practice.

Connecting Sound Card Audio to the Transceiver

On the sound card side, things are straightforward. If your PC has Line In and Line Out jacks you should use them as your first choice. If, as on most laptops, you only have Microphone In and Speaker Out, use those instead. If using the PC's Speaker-Out signal, its level often needs to be attenuated before it can be applied to the microphone input of your transceiver.

The advantage of using your sound card's Line In and Line Out jacks is that they have separately settable volume levels and will not interfere with the volume levels you may be using for a headset or an external set of speakers. Also, the Line Out signal, being of a lower amplitude, might be usable without requiring attenuation.

Keying your Transceiver

In addition to cables for routing audio signals between the PC and your radio, you need to provide a method for keying your radio's transmitter. This is typically done either by interfacing a signal on your PC's serial port (RTS or DTR), via a level shifting circuit, to the radio's transmit keying input, or by using your radio's VOX (voice operated transmit). However, VOX is not normally found on VHF/UHF radios.

Depending upon the type of PC you have, even serial port triggered transmission may present a problem. Older PCs and laptops almost always came with a serial port that could be used. Recent vintage PCs, on the other hand, generally lack serial ports in favor of providing one or more USB ports.

If your PC or laptop does not have a serial port you could use a USB to serial port adapter. If you don't have an adapter you could buy one. If you are going to have to spend some money on an adapter, though, you might instead want to consider purchasing one of the commercial interfaces discussed in Chapter 7 that already include a USB port.

You may have to do some improvising to connect a PTT signal to your transceiver. Depending upon your radio, one of the following solutions may be the way to go:

- Connect the PTT signal to an auxiliary connector at the rear of your radio. If you select this approach, you may also be able to connect the sound card signals to the same connector. Consult the owner's manual for your radio.
- Connect to the radio's microphone input jack. If you do this you can also connect the sound card's Speaker or Line-Out signal to the microphone input pins on the jack.

Establishing a Repeater Node

There are many similarities and some significant differences when setting up simplex nodes and repeater nodes. Repeater-unique issues and settings are explicitly mentioned and covered throughout this book as a part of individual topic discussions. In particular, Chapters 10 and 11 should be reviewed before implementing a repeater node. Also, you should never attempt to establish an EchoLink node on an existing repeater without first procuring permission from the repeater owner or administrator.

PC to Transceiver Interfaces

If desired, you could put together some circuitry and cabling to accomplish a PC-to-Transceiver interface that meets the above-described requirements. Or, if you already have an interface that you have been using with other digital modes (PSK, RTTY, etc), you might be able to use that same hardware for your EchoLink PC-to-transceiver interface.

For most of us, purchasing a proven interface is probably the way to go. It's much simpler and will save a lot of experimentation and troubleshooting.

Via either the **Link Setup Wizard** or the **Sysop Setup** windows, the EchoLink software supports two main types of interfaces:
- ASCII-controlled interfaces designed expressly for EchoLink.
- General purpose digital mode PC interfaces

Interfaces Specifically Designed for EchoLink: In 2010 when this was being written, I was able to locate three ASCII-controlled interfaces:
- *ULI* from WB2REM/G4CDY at **http://www.ilinkboards.com/**
- *VA3TO* available at **http://www.ilinkca.com/**
- *EI-151* from KJ6ZD at **http://www.kj6zd.net/**

Both the WB2REM *ULI* interface and VA3TO boards are mentioned in the EchoLink help files. All three of these interfaces accept ASCII commands via their serial port to key transmission. The WB2REM / G4CDY *ULI* interface is discussed more fully in Chapter 7. While these boards were expressly designed to support EchoLink, they also support most all other modes of digital operation that you might want to operate.

EchoLink Specific Interface Features

EchoLink specific interfaces generally include a number of features not contained in their general-purpose cousins. The main features to look for are:

- A maximum transmit timer that cuts off transmission after a programmable selectable period of time.

- Hardware detection of DTMF tones for improving the reliability of tone detection.

- An internal sound mixer for combining audio from an auxiliary control transceiver, useful for implementing remote control DTMF commands.

- Ability to detect and level-shift a link transceiver's COS (carrier sense signal) passing it to the EchoLink software.

- An ASCII interface to the PC for controlling transmission and receiving DTMF tones from the interface. Other ASCII functions are also generally included.

While the EchoLink software has its own user settable maximum transmit timer, if it times out, the station transmitting is automatically disconnected from your node. This is a bit of overkill, and perhaps rude. Implementing a maximum transmit timer in the interface hardware allows an excessively long transmission to be cut off without disconnecting the user from the node; resulting in an operation quite similar to the familiar "alligator" maximum transmit timeout found on most FM repeaters.

General Purpose Digital Mode Interfaces: A variety of general purpose digital mode interfaces are manufactured by several different companies. These interfaces are designed to support virtually all digital modes: PSK, MFSK, SSTV, RTTY, AMTOR, packet and many others, including EchoLink. A variety of interfaces are available from the following manufacturers:
- Various *RIGblaster* models from West Mountain Radio
- The *SignaLink USB* from Tigertronix
- Various Sound Card Interface models from MFJ

The *RIGblaster Plus II* from West Mountain Radio and the *SignaLink USB* from Tigertronix are discussed more fully in Chapter 7.

Chapter 7: Commercial PC Interfaces

As examples of alternative methods for interfacing transceivers to a PC, in this chapter we take a detailed look at implementing an EchoLink node with three different commercially made interfaces:
- West Mountain Radio's *RIGblaster Plus II*,
- WB2REM's *ULI*
- TigerTronics *SignaLink USB*.

The fact that they are discussed in detail in no way implies that we think they are the only solutions you should use. Like most things, the selection of a PC-to-Radio interface is a personal choice, depending upon the radio you have, your individual requirements, and your budget.

Chapter 7, Section 1: RIGblaster Plus II Interface

West Mountain Radio's *RIGblaster Plus II* is a good example of a general-purpose digital mode interface. It's a reasonably priced middle-of-the-road interface. There are more complex and feature-rich interfaces, and likewise simpler bare-bones interfaces are also available. Most general purpose interfaces can be used for EchoLink operation as well as being compatible with most of the other digital modes: PSK, MFSK, SSTV, RTTY, AMTOR, SSTV, WEFAX etc. At the time this guide was written, the list price of the *RIGblaster Plus II* on the **www.westmountainradio.com** web site was $159.95 with free shipping. This interface is also available from most amateur radio retailers and is worth considering if you need a USB interface, and don't intend on using a COS (carrier sense signal).

As a rule, most general purpose interfaces provide isolation for the microphone sound card signal and have some method for keying the transmitter. The differences between interfaces are in how this is performed, and how easily the interface can be adapted to a variety of radios. Many models include features simplifying the connection of optional equipment, adjusting audio signal levels, providing for the connection of auxiliary speakers and microphones and for a variety of status indicators.

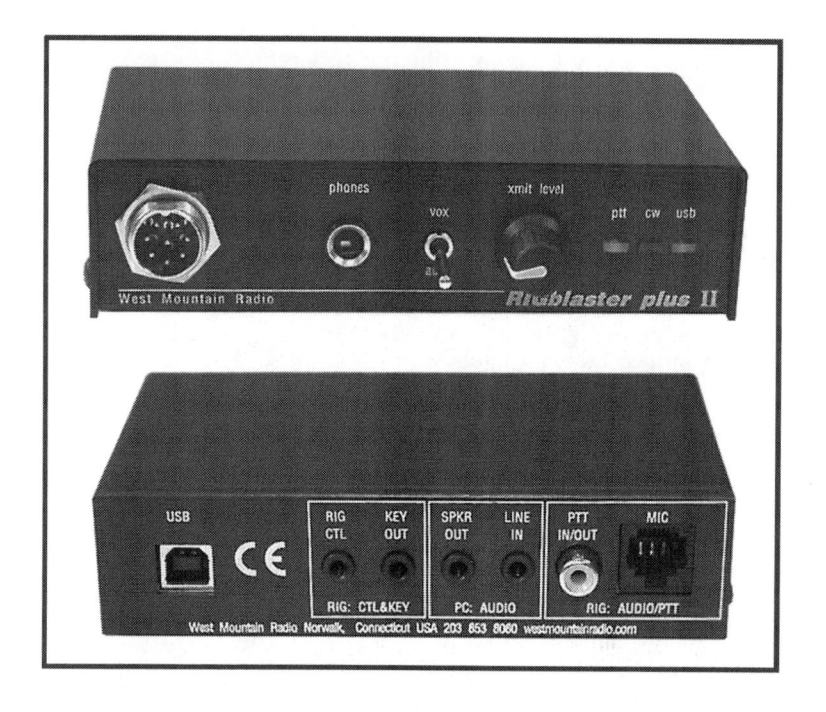

Distinctive Features of the RIGblaster Plus II

The *RIGblaster Plus II* is representative of the level of EchoLink support provided by general purpose digital interfaces. *RIGblasters*, depending on the model, support a number of useful and desirable features when used as a station's general purpose PC-to-transceiver interface. In this book, though, we are primarily concerned with features that support the creation of an EchoLink RF node. The following list highlights features useful for supporting EchoLink operation.

- Transformer isolated soundcard to transceiver transmit audio.
- USB operated, convenient for PCs without a serial port.
- Powered by the USB cable, eliminating a "wall wart"
- Status LED indicating when the PC recognizes a connection to the RIGblaster's USB port.
- Status LED indicating when the link transceiver is transmitting

- Simplified cabling interconnection to most amateur radios
- CAT / CI control of amateur transceivers, which allows changing frequency and making other adjustments to the link transceiver via a compatible PC program.

RIGblaster Plus II Interface Cables

The *RIGblaster Plus II* is designed to work through a transceiver's microphone connector. This greatly reduces the number of cable configurations that must be supported. Connecting to the radio's microphone input avoids cabling and sound quality issues that may arise when connecting to a radio's auxiliary or data connectors. Output audio from the link transceiver is routed directly to the computer's sound card input jack, without passing through the interface.

A full complement of cables and "Instant Setup Connectors, ISCs" are supplied with the interface. This combination of cables and ISCs should be sufficient for interfacing to most amateur radio transceivers. West Mountain Radio says it's compatible with over 2000 different radios. For old or unusual radios that may not be supported by the ISCs and cabling provided with the interface, West Mountain Radio advises checking their support page for additional cabling information.

The following cabling supplies are supplied with the interface:
- Six ISC connectors. These are "jumper headers" with a small printed circuit board that routes signal connections between various pins on the header. Depending on the radio you have, one of the six ISC connectors is plugged into the circuit board inside the interface. This simplified "re-wiring" ability allows the same set of external cables to be used with a large variety of different radios.
- One RJ-45 to 8-pin screw-on microphone interconnect cable.
- Two 1/8" phone plug cables for connecting to your computer's sound card.
- One USB cable for communicating with the computer.

Power Supply Connection:
The *RIGblaster Plus II* is powered via its USB interface cable, so no additional power supply connections are required.

RIGblaster Plus II Functional Block Description

From a functional standpoint, when used for EchoLink operation the *RIGblaster Plus II* is typically connected between the PC and your transceiver as shown below.

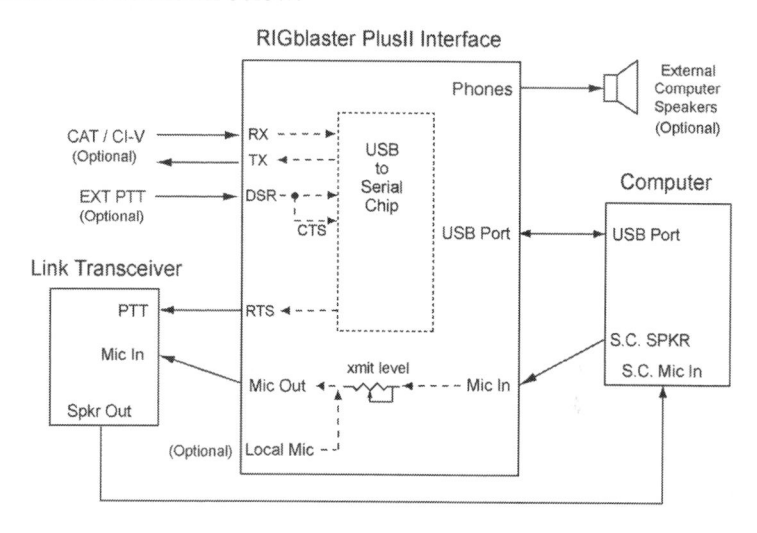

Typical RIGblaster Plus II Interface Interconnect Diagram

One of the key components within the *RIGblaster Plus II* is the USB-to-Serial conversion chip. This circuit converts USB formatted data received from the computer into the equivalent standard RS-232 style serial port data communication and controls. The above diagram illustrates the basic functionality. The normal serial **RX** and **TX** communication lines are dedicated to CAT / CI-V control of compatible transceivers, which could be useful for optional control of the radio, but is not essential for EchoLink operation.

Transmit keying of the link transceiver is performed by ASCII commands received from the EchoLink software on the PC. These transmit commands are routed via the PC's USB port to the

RIGblaster Plus II's USB chip, resulting in level changes to the serial port **RTS** pin. The resulting **PTT** signal is routed to the link transceiver's microphone input connector. It's also possible to key the transceiver from a locally connected microphone or the **EXT PTT** jack. While an optionally connected microphone can be used for communication over the link transceiver, it cannot be used for Internet communication.

As shown in the above diagram, the *RIGblaster Plus II's* front-panel has a variable control, **xmit level**, used to adjust the level of the speaker out signal coming from the PC's sound card. This signal is then routed to the link transceiver's microphone input line.

If desired, external computer speakers can be attached to the *RIGblaster Plus II*. This can be handy for monitoring audio going to and from the Internet, especially useful during set up and testing. As the volume settings for the link transceiver, computer and *RIGblaster* have to be adjusted for proper transmission levels over the transceiver and Internet, any externally connected speakers should have their own amplifier and volume control.

The link transceiver's Speaker Out audio is cabled directly to the computer's sound card Mic In or Line In inputs. It is not connected to the *RIGblaster* interface. Depending upon your radio, you might find it necessary to attenuate this signal by using a resistive divider network. Exact values are not critical; it's the resistance ratio that determines the amount of attenuation. The following circuit provides a 10:1 reduction in amplitude.

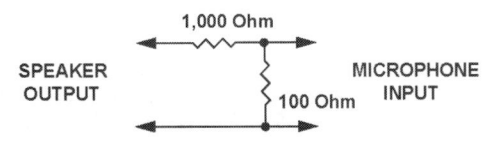

```
                    1,000 Ohm
                 ◄───/\/\/\───►
   SPEAKER         │       │      MICROPHONE
   OUTPUT          ⌇ 100 Ohm     INPUT
                 ◄───────────►
```

Speaker Audio Attenuator

RIGblaster Plus II Installation and Set Up

Installation of the *RIGblaster Plus II* involves selecting and installing the correct ISC to use with your radio, plugging in the external cables and installing the *RIGblaster's* USB driver. Refer to the manual supplied with the interface for instructions on performing these tasks.

For EchoLink operation, place the **vox/auto** switch in the auto position.

Sysop Settings for use with the RIGblaster Plus II

A few Sysop settings must be configured for the EchoLink software to work properly with the *RIGblaster Plus II* interface. These same settings should also work with most general-purpose interfaces.

Under the **Tools** menu, select **Sysop Settings**, and on the window that opens, select the **RX Ctrl** tab. Since the *RIGblaster Plus II* does not have provisions for connecting a COS signal, select **VOX**, which instructs EchoLink to use its internal software based VOX mechanism for detecting when a valid audio signal is being received. A little later in this section you will also need to adjust the VOX threshold level slider found under the EchoLink screen's audio-level meter. Additionally, as initial presets, set the remainder of the fields as shown in the VOX set up screen imagess below.

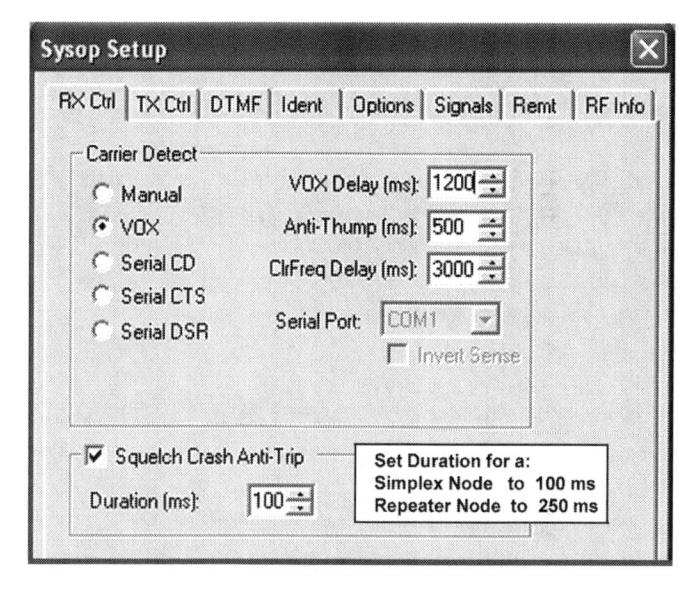

Receive Control settings for use with VOX

The **Squelch Crash Anti-Trip** setting prevents unwanted noises from being transmitted to the Internet. For simplex node operation, the **Duration** can be set to 100 ms, for repeater nodes set it to 250 ms.

Next select the **TX Ctrl** tab, and select **RTS** as the **PTT Activation** method. This informs EchoLink that the *RIGblaster Plus II* interface will be using the **RTS** serial port signal to trigger transmission of the link transceiver.

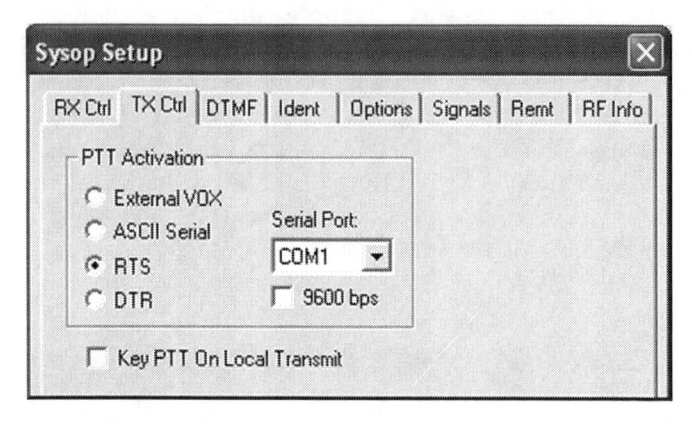

Transmit Control settings for use with RIGblaster Plus II

Also select the **COM** port to be used with the *RIGblaster* and leave **9600 bps** unchecked. Refer to the *RIGblaster Plus II's* Owners Manual for instructions on how to determine which COM port the Windows operating system assigned during installation of the *RIGblaster's* USB software driver.

Setting Received Audio Drive Level from the Transceiver

Received audio is wired directly from your transceiver to the sound card's microphone (or line-in) jack and the *RIGblaster Plus II* has no influence on it. Adjustment of the received audio level from the Link Transceiver is accomplished using a combination of the radio's output volume control and your computer's recording level control.

RX Drive Level Procedure:
1. Adjust the link transceiver's volume control to produce normal audio volume while monitoring received signals over its speaker. Then open your computer's recording volume control window, initially setting the slider to mid range.

To open the computer's volume control, select the **Adjust Sound Device** entry found under the **Tools** pull-down menu. Select the **Recording** option, which brings up the **Recording Control** volume adjustment window.

If the **Recording Control** window includes controls for **Wave Out** or **Wave Out Mix**, disable these as they can interfere with reliable DTMF operation.

2a. If you are setting up your link transceiver as an EchoLink simplex node, make the following adjustment while using another radio to transmit on the link's simplex frequency.

2b. If you are setting up your link transceiver as a repeater Link node, make the following adjustment while "normal" level audio is being transmitted from the repeater.

3. While the link transceiver is receiving a signal, observe EchoLink's audio-level meter. Using your computer's **Microphone** recording level control, adjust the slider so that audio peaks are starting to overdrive the meter (a yellow bar indicates overdrive), then back off until audio peaks only overdrive the meter on loud audio peaks. If you don't have enough range, you might need to open the **Advanced** window of the **Recording Control** screen and select the **20db boost** option.

Setting EchoLink's Received Audio VOX Threshold

Once the receive drive level has been set, we can set EchoLink's received signal VOX threshold level. It's important to set the VOX threshold level sufficiently high to prevent undesired EchoLink transmissions due to receiver noise pickup. On the other hand it needs to be sensitive enough to reliably trigger transmission when receiving weak signals from the radio.

When in Sysop mode, an adjustment slider is displayed under the audio-level meter. When the audio level exceeds the VOX threshold level set by the slider, EchoLink will transport audio being received from the link transceiver to the Internet. It's not necessary to be in an actual QSO with a remote station to perform these adjustments; monitoring local link transceiver activity is sufficient.

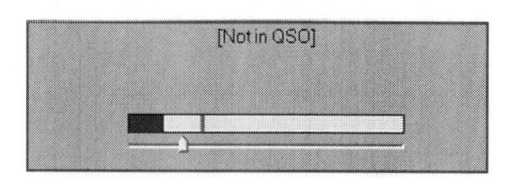

Setting Received Audio VOX Level

Setting Simplex Node VOX level:
Setting the VOX trigger level for Simplex node operation is rather simple, just set it about halfway between an un-modulated carrier level and the level seen when receiving a fully modulated carrier. While operating, you can increase the sensitivity if transmission is not being triggered reliably on weak signals, or decrease the sensitivity if experiencing false triggering.

Setting Repeater Node VOX level:
Setting the VOX trigger level is a bit more critical for repeater node operation. While monitoring your repeater, make note of the difference in signal levels when receiving voice, repeater IDs and courtesy tones. Ideally repeater ID and courtesy tones will produce lower signal levels than voice.

If there is a pronounced difference between voice and the repeater's IDs and courtesy tones, it should be possible to set the VOX threshold between these levels and prevent unwanted signals from being transmitted over the EchoLink network. This will not, however, eliminate ID's that occur concurrently while a user is transmitting.

If there is no significant difference in the levels, perhaps the repeater controller can be adjusted to reduce the volumes of the repeater ID and courtesy tone. Failing that, you will need to consider alternative methods for keying transmission to the EchoLink network. Refer to Chapter 10 for a discussion of using CTCSS or repeater co-location as methods of solving the problem.

For repeater operation, make sure the **Squelch Crash Anti-Trip** setting on the Sysop **Rx Ctrl** tab is initially set to 250 ms, which is a typical setting. If the VOX trigger level has been set correctly and you are still having trouble with courtesy tones or squelch crashes triggering transmission, this delay can be increased.

For detailed instructions on setting the VOX level, refer to the **Repeater Linking Tips** section of the EchoLink help files.

When you think you have the VOX set properly, verify that when you connect to the EchoLink Test Server, that after hearing the welcome message that the repeater stops transmitting and remains idle. Recording and playing back your own voice should also operate normally and not result in unnecessary additional transmissions from either the Test Server or your station.

Setting Audio Drive Level to the Link Transceiver

Adjustment of the transmit audio level going to the link transceiver is accomplished using a combination of the *RIGblaster's* front-panel **xmit level** control and the computer system's output Volume Level control. The computer system's Volume Control is used as a master overall range setting. Once set, you can generally forget about it and use the *RIGblaster's* **xmit level** control to make any needed transmitted audio level adjustments.

VOX Adjustment and TX Drive Level Procedure:
1. Preset the *RIGblaster's* **xmit level** control knob to about the middle of its range.

2. Using the **Adjust Sound Device** setting found under the **Tools** pull-down menu. Select the **Playback** option, which brings up the **Volume Control** adjustment window.

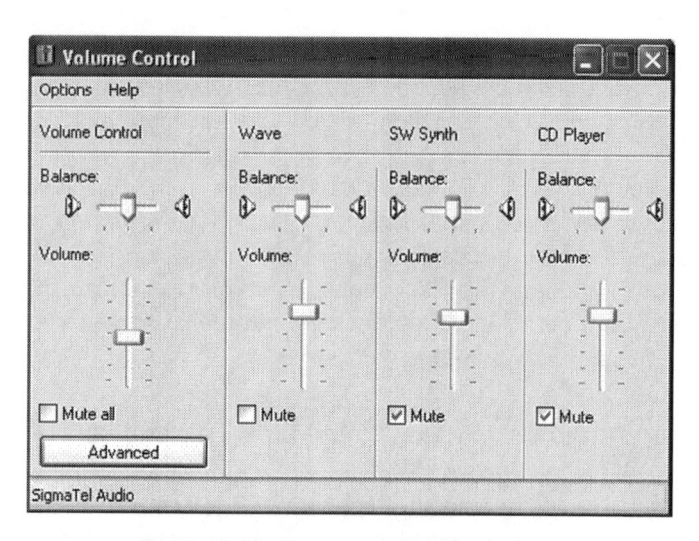

Setting the Transmit Drive Level

3. Make sure **Volume Control** and **Wave** are not muted. If desired you can mute other audio sources to prevent any bleed-over from them. Initially set **Wave** to about 2/3 of its maximum setting.

4. Now we can set the transmitted audio level using EchoLink's Test Server greeting. To verify the drive level you are setting, you will need to monitor the output of your link transceiver with a second transceiver.

   Note: Actually the best way to set this is by using a deviation meter or service monitor. If you have one of these pieces of equipment and know how to set deviation you can disregard using a second transceiver as described below.

   With a second transceiver tuned to the output of your link transceiver, connect to EchoLink's Test Server. To complete this procedure you will most likely have to connect and disconnect multiple times to generate audio for making the adjustment.

   While monitoring the Test Server's greeting, using the *RIGblaster's* **xmit level** control increase the volume to the point where it no longer increases as the control is turned further. You have now reached the point of maximum deviation and some audio distortion may be noted. Now reduce the **xmit level** setting to the point where the Test Server's greeting begins to decrease in volume and then decrease it just a little bit more. The link transceiver's transmit drive level should now be set.

Congratulations! You now have your *RIGblaster Pro II* configured for EchoLink operation. If you have not yet already done so, make sure that all the settings found under the various tabs of the Sysop Setup window are properly set.

Now that the *RIGblaster Pro II* is configured and receive and transmit levels have been set, you should verify the VOX threshold settings. Setting EchoLink's VOX threshold slider for proper detection of voice signals from your link transceiver, while eliminating other signals is especially important.

If setting up a Repeater Link, pay particular attention to the **Anti-Thump** and **Squelch Crash Anti-Trip** settings found on the **RX Ctrl** tab of the **Sysop Setup** window. Information for making these adjustments is found in the "RX Ctrl Tab" section of Chapter 9.

.

Chapter 7, Section 2: **WB2REM/G4CDY ULI Interface**

Designed primarily for EchoLink applications, the WB2REM / G4CDY *ULI*, "Ultimate Linking Interface" includes a number of useful features targeted for EchoLink operation which are not found on general-purpose digital mode interfaces. On the other hand, it can also be used as a general-purpose PC-to-transceiver interface for all popular digital modes of operation: PSK, MFSK, SSTV, RTTY, AMTOR, SSTV, WEFAX etc.

Distinctive Features of the WB2REM / G4CDY ULI Interface

The *ULI* and its earlier iterations were designed especially for EchoLink operation. It has been around long enough that it can be considered somewhat of a standard for setting up EchoLink nodes. Provisions for it are specifically included in the EchoLink Sysop Setup menus. The following list highlights some of the unique features incorporated into the *ULI*.

- Ability to use a COR / COS signal from the link receiver to positively key transmission to the Internet.
- Ability to adjust the COR / COS trigger level (settable via a program).
- Ability to combine audio signals from four different inputs: an auxiliary control receiver, the main link transceiver, the sound card and a locally connected microphone.
- Ability to mix audio signals simplifies using a second receiver for auxiliary remote DTMF control of the node.
- Ability for hardware detection of DTMF tones for improved detection reliability.
- Ability to detect DTMF controls coming from the Internet.
- Ability to key transmission using the RTS / CTS serial port signals from a PC.
- Ability to receive an ASCII control character "**T**" from the PC as an alternative method of keying transmission.
- Ability to set a maximum transmit time to prevent long-winded EchoLink users or software / computer failures from keying the link transceiver for excessive periods.
- Ability to use an internal watchdog timer and PC hardware reset line to automatically restart a computer should it crash.
- Ability to reduce any residual high frequency noise coming from the sound card via a built in 300~3000 Hz audio filter.

The WB2REM / G4CDY *ULI* is an interface worth considering if you intend to make use of the COR / COS, carrier sense signal on your link transceiver. Additionally, since the interface has an audio mixer capable of mixing DTMF control signals from an auxiliary receiver, remote control is simplified. Depending on your installation, some of the other features listed above might also be helpful. The maximum transmit time timer, which automatically cuts off transmission, is nice to have because it can prevent EchoLink users from suddenly being disconnected from your node if they get a little longwinded. At the time this guide was written, the list price of the *ULI* on the **www.ilinkboards.com** web site was $134 plus shipping.

ULI Interface Cables and Power Source

The *ULI* does not come with any interface cables, you have to provide your own.

You have two choices when connecting to your link transceiver:
1. Via the *ULI's* RJ45 connector, which has all necessary signals, or
2. Via individual 1/8" (3.5mm) mono phone plug cables for connecting to your radio's speaker out, Mic in, PTT and to a COR / COS take off point if you will be using that.

There are two choices when using the *ULI's* RJ-45 jack:
1. A few radios like the IC-706 can be directly connected with a straight thru RJ-45 cable. (Check the documentation for your particular radio for cabling / signal compatibility.)
2. Most other radios will require breaking out the RJ-45 cable and changing / rewiring some of the signals. If that is the case, you might be better off using the individual phone plug cable method of connecting to your radio described below.

(Note 1: For general simplex node testing using my IC-7000, I was able to use a RJ-45 cable after clipping (opening) pin 3, which was not compatible with the *ULI's* RJ-45 pin 3 *Audio Input* requirement. Then using a single 1/8" phone plug cable I connected the IC-7000's aux speaker out signal to the *ULI's* **Main Rx** audio in jack. Similar changes might be required for your transceiver.)

(Note 2: For setting up a permanent repeater node using a surplus GE MVS mobile radio, I used the phone plug method described directly below.)

Cables for the phone plug method of connecting your radio:
Three shielded cables with a 1/8" (3.5mm) mono phone plug on one end, and bare wire for solder termination to your equipment on the other. You can get these cables from Radio Shack or elsewhere, if you don't already have them. They can be purchased from Mouser Electronics, at $2.20 each, Mouser part number 172-2136. Two of these cables are used for PTT and microphone inputs to the link transceiver. A third is needed if you plan on using a COR / COS connection to your transceiver.

An additional shielded cable with 1/8" (3.5mm) mono phone plugs on both ends is required for connecting the output audio from your transceiver to the *ULI*. This cable can be purchased from Mouser Electronics, at $3.15 each, Mouser part number 172-2137.

Mouser parts can be bought on-line at **www.mouser.com**.

Cables required for connecting to the PC's Sound Card:
Two additional shielded cables with 1/8" (3.5mm) mono phone plugs on both ends are required for connecting the *ULI* to your computer's sound card Mic in and Speaker out jacks. These are the same as the Mouser 172-2137 cables mentioned directly above.

Power Supply Connection:
In addition to the cables described above, you also need a regulated and filtered DC source for powering the *ULI*. The voltage can be anything between 9 and 15 volts. This could be from the power supply running your transceiver, the transceiver itself, or from an AC wall adapter. The *ULI* comes with a power cable for connecting your power source to the *ULI*.

ULI Functional Block Description

From a functional standpoint, the *ULI* is typically connected between the PC and your transceiver as shown below.

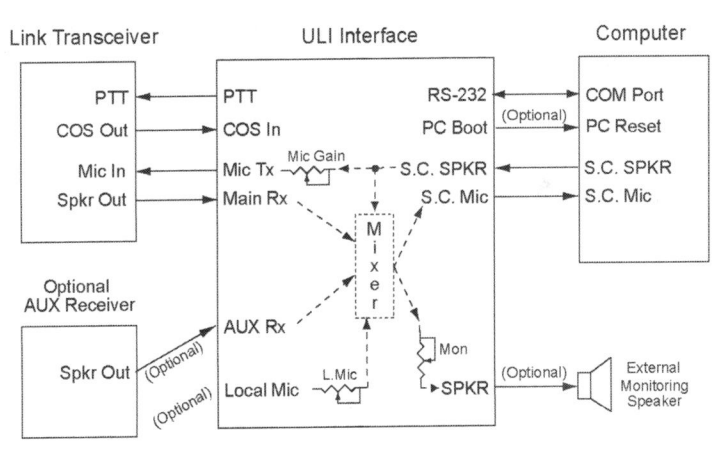

Typical ULI Interface Interconnect Diagram

The presence of received signals on the link transceiver can be communicated using COS, the carrier detect signal, which is passed to the PC via the **CD** pin of the *ULI's* RS-232 serial port. (If not using COS, EchoLink's received signal VOX mechanism can be used.)

Transmit keying of the link transceiver is performed by commands received from the EchoLink software residing on the PC. The transmit command can be routed to the Link Transceiver via an ASCII "T" command sent via the PC's serial port, or via the serial port's **RTS** pin. Upon receipt of either signal, a microcontroller within the *ULI* will activate the PTT line going to the link transceiver.

The *ULI's* **Setup Software**, uses the serial port is to configure several user adjustable parameters within the *ULI's* hardware. If your computer does not have a serial COM port, you will need a USB-to-serial adaptor.

An audio mixer within the *ULI* can be used to combine audio from: the link transceiver, an optional auxiliary control receiver, and from a microphone connected directly to the *ULI*. The auxiliary input into the mixer simplifies implementing the auxiliary receiver method of DTMF remote control.

The *ULI* contains a DTMF decoder chip, which is connected to the output of the audio mixer. Connection to the output of the mixer allows DTMF control signals to be received from either the main link transceiver or the auxiliary receiver, or optionally even from the Internet. A microcontroller within the *ULI* monitors for received DTMF signals and transmits corresponding digital DTMF codes to the EchoLink software via the serial port.

If desired, an external monitoring speaker can be attached to the *ULI*. This can be handy for monitoring all audio going to and from the Internet, especially useful during set up and testing. Unlike general-purpose interfaces like the *RIGblaster*, the ULI comes equipped with a dedicated volume control and amplifier for use with an externally connected speaker.

As shown in the preceding diagram, the *ULI's* front panel has two variable potentiometers. The one labeled **MIC GAIN** is used to adjust the gain of the Speaker Out signal coming from the PC's sound card. The one-labeled **MON LEVEL** adjusts the volume of an optionally connected external speaker. In addition to these two front panel controls, a third trimmer potentiometer mounted on the *ULI's* circuit board can be used to adjust the audio level coming from an optionally connected microphone.

A feature unique to the *ULI* is the **PC Boot** signal, which can be used to restart the computer should it crash. Internal to the *ULI* there is a watchdog timer that expects to see a periodic signal from the PC. If this signal fails to arrive, the **PC Boot** signal can be used to restart the computer. This can help improve the reliability of unattended EchoLink node operation. Often desktop computers have a header on the motherboard that connects to the reset button, providing easy access to its internal restart circuitry. If your computer lacks such a header and still want to use this feature, you will need to locate a place to tap into the computer's circuitry. The **PC Boot** signal is available on a 2-pin header, **JP2**, on the *ULI's* circuit board.

WB2REM / G4CDY ULI Installation and Set Up

ULI installation documentation is available for download at **http://www.ilinkboards.com/ULI1.2Files.zip**. Create a new folder on your computer, naming it **uliboardsetup**. Download the zip file, copying it into the **uliboardsetup** folder. Unzip the **ULI1.2Files.zip** file into the same folder.

Unzipping the file creates a folder called **ULI1.2Files**, which contains six sub-folders. Unless you are building your *ULI* as a kit, you will most likely not need the information contained in the **Enclosure**, **Kit Instructions** or **Parts List** folders.

Set up instructions and other information is found in the **User Docs** folder. The first document you should open and print is the **ULI BOARD WIRING INSTRUCTIONS(1.2).doc** file, which contains instructions for connecting your transceiver and computer to the *ULI*.

Additional files should be opened and printed as needed:

ULI_CONNECTIONS.pdf   Explains functions of all the connectors and includes a component placement diagram and a schematic of the main board.

uli_case1.pdf   Describes the function of the front panel LEDs and switches.

Uli_software.pdf   Explains how to run the **ULI Board Setup** program. The *ULI* can initially be used with the defaults programmed into it at the factory. If desired, the set up program can be used to modify these default settings. The following list summarizes the parameters that can be set, and the default values programmed by the factory.

- **TX Timeout Secs**   (max transmit time out – **180 sec**)
- **COR Threshold V**   (COR detection voltage – **2.08 V**)
- **PC Reboot Timer**   (max watch dog timer – **120** sec)
- **7 User Programmable CAT functions** (**none** programmed)

Important: Ignore the **Installing the Software** section of the **Uli_software.pdf** document, as it is no longer correct. Instead, to install the **ULI Board Setup** program, unzip the **ULISetupSoftware.zip** file. This creates a sequence of five folders all named **ClickHere** that you will need to click through before arriving at the actual installation software. The final file is called **ULI SETUP PROG V0.4.msi**. Click on it to install the program.

When you initially start the **ULI Board Setup** program, first select the **COM** port you are using, then it's a good idea to read and store to a file of your choosing the values as originally shipped from the factory. This is accomplished by pressing the **READ ULI** button to download the *ULI's* current settings, and then by using the **File** save pull down button as usual for Windows programs.

Sysop Settings for use with the ULI

A few Sysop settings must be configured for the EchoLink software to work properly with the *ULI* interface.

Under the **Tools** menu, select **Sysop Settings**, and on the window that opens, select the **RX Ctrl** tab. If you are going to be using your link transceiver's COS signal, select **Serial CD** as the received **Carrier Detect** mechanism. Otherwise select **VOX**, which instructs EchoLink to use its internal software-based VOX mechanism to detect when a valid audio signal is being received from the link transceiver. If you selected **VOX**, a little later in this section you will also need to adjust the VOX level slider found under the audio level meter. Additionally, set the COM port you will be using and as initial presets set the remainder of the fields as shown in either the COS or VOX set up screen images below.

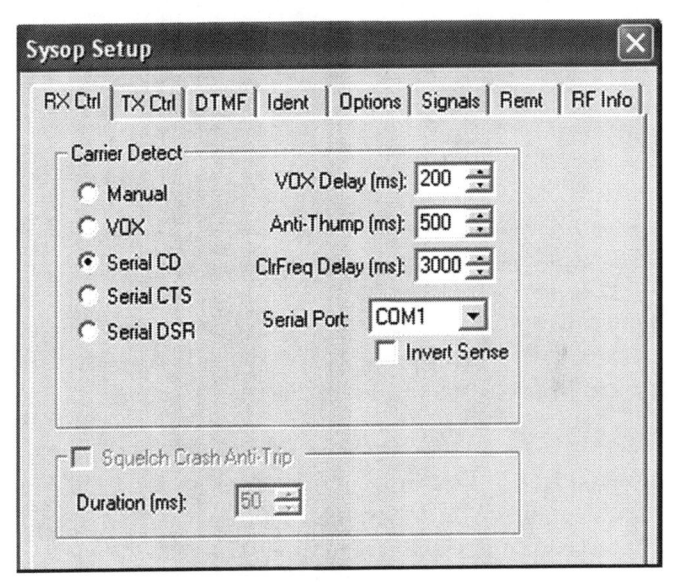

Receive Control settings for use with COS

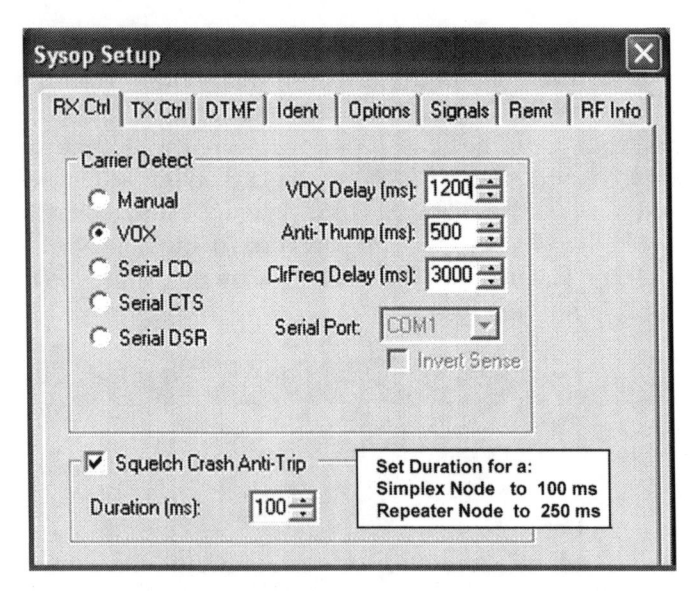

Receive Control settings for use with VOX

The **Squelch Crash Anti-Trip** setting prevents unwanted noises from being transmitted to the Internet. For simplex node operation, the **Duration** can be set to 100 ms, for repeater nodes set it to 250 ms.

Next select the **TX Ctrl** tab, and then select **ASCII Serial** as the **PTT Activation** method. This informs EchoLink that the *ULI* interface will be sending an ASCII control character to trigger transmission of the link transceiver when audio is received from the Internet.

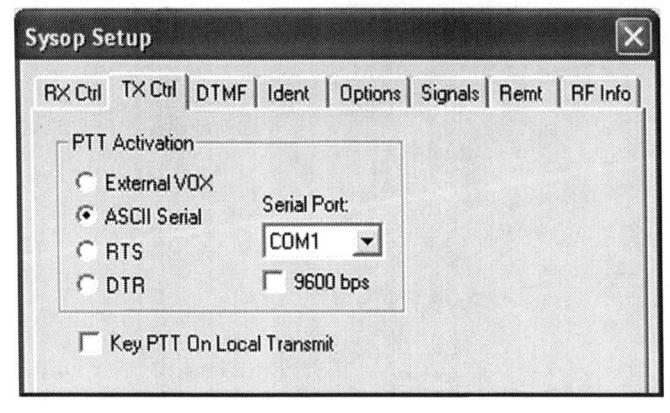

Transmit Control settings for use with the ULI

Also select the **COM** port to be used with the *ULI* and leave **9600 bps** unchecked, as the ULI operates at 1200 bps.

Setting Received Audio Drive Level from the Transceiver

Adjustment of the received audio level from the link transceiver is accomplished using a combination of the radio's output volume control and your computer's recording level control.

RX Drive Level Procedure:
1. If the *ULI* is connected to the link transceiver's speaker output, adjust the link transceiver's volume control to produce normal volume audio. Then open your computer's recording volume control, initially setting the slider to mid range.

   To open the computer's volume control, select the **Adjust Sound Device** entry found under the **Tools** pull-down menu. Then select the **Recording** option, which brings up the **Recording Control** volume adjustment window.

   If the **Recording Control** window includes controls for **Wave Out** or **Wave Out Mix**, disable these as they can interfere with reliable DTMF operation.

2a. If you are setting up your link transceiver as an EchoLink simplex node, make the following adjustment while using another radio to transmit on the link's simplex frequency.

2b. If you are setting up your link transceiver as a repeater link node, make the following adjustment while "normal" level audio is being transmitted from the repeater.

3. While the Link Transceiver is receiving a signal, observe EchoLink's audio level meter. Using the computer's **Microphone** recording level control, adjust the slider so that audio peaks are starting to overdrive the meter (a yellow bar indicates overdrive), then back off until audio peaks only overdrive the meter on loud audio peaks. If you don't have enough range, you might need to open the **Advanced** window of the **Recording Control** screen and select the **20db boost** option.

Setting EchoLink's Received Audio VOX Level

If you are using COS instead of the EchoLink's VOX method of carrier detection, you can skip this section.

If using VOX, once the receive drive level has been set, we can set EchoLink's received signal VOX trigger level. It's important to set the VOX threshold level sufficiently high to prevent undesired EchoLink transmissions due to receiver noise pickup. On the other hand it needs to be sensitive enough to reliably trigger transmission when receiving weak signals from the radio.

When in Sysop mode, an adjustment slider is displayed under the audio-level meter. When the audio level exceeds the VOX threshold level set by the slider, EchoLink will transport audio being received from the link transceiver to the Internet. It's not necessary to be in an actual QSO with a remote station to perform these adjustments; monitoring local link transceiver activity is sufficient.

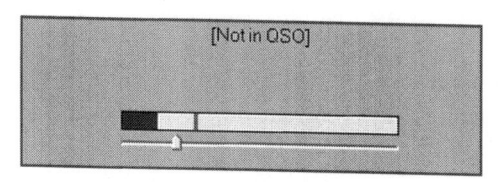

Setting Received Audio VOX Level

Setting Simplex Node VOX level:
Setting the VOX trigger level for Simplex node operation is rather simple, just set it about halfway between an un-modulated carrier level and the level seen when receiving a fully modulated carrier. While operating, you can increase the sensitivity if transmission is not being triggered reliably on weak signals, or decrease the sensitivity if experiencing false triggering.

Setting Repeater Node VOX level:
Setting the VOX trigger level is a bit more critical for repeater node operation. While monitoring your repeater, make note of the difference in signal levels when receiving voice, repeater IDs and courtesy tones. Ideally repeater ID and courtesy tones will produce lower level signals than voice.

If there is a pronounced difference between voice and the repeater's IDs and courtesy tones, it should be possible to set the VOX threshold between these levels and prevent unwanted signals from being transmitted over the EchoLink network. This will not, however, eliminate ID's that occur concurrently while a user is transmitting.

If there is no significant difference in the levels, perhaps the repeater controller can be adjusted to reduce the volumes of the repeater ID and courtesy tone. Failing that, you will need to consider alternative methods for keying transmission to the EchoLink network. Refer to Chapter 10 for a discussion of using CTCSS or repeater co-location as methods of solving the problem.

For repeater operation, make sure the **Squelch Crash Anti-Trip** setting on the Sysop **Rx Ctrl** tab is initially set to 250 ms, which is a typical setting. If the VOX trigger level has been set correctly and you are still having trouble with courtesy tones or squelch crashes triggering transmission, this delay can be increased.

For detailed instructions on setting the VOX level, refer to the **Repeater Linking Tips** section of the EchoLink help files.

When you think you have the VOX set properly, verify that when you connect to the EchoLink Test Server, that after hearing the welcome message that the repeater stops transmitting and remains idle. Recording and playing back your own voice should also operate normally and not result in unnecessary additional transmissions from either the Test Server or your station.

Setting Audio Drive Level to the Link Transceiver

Adjustment of the transmit audio level going to the Link Transceiver is accomplished using a combination of the *ULI's* front-panel **MIC GAIN** control and the computer system's output Volume Level control. The computer system's Volume Level control is used as a master overall range setting. Once set, you can generally forget about it and use the *ULI's* **MIC GAIN** control to make any needed transmitted audio level adjustments.

VOX Adjustment and TX Drive Level Procedure:

1. Preset the *ULI's* **MIC GAIN** control knob to about mid range.

2. Using the **Adjust Sound Device** setting found under the **Tools** pull-down menu. Select the **Playback** option, which brings up the **Volume Control** adjustment window.

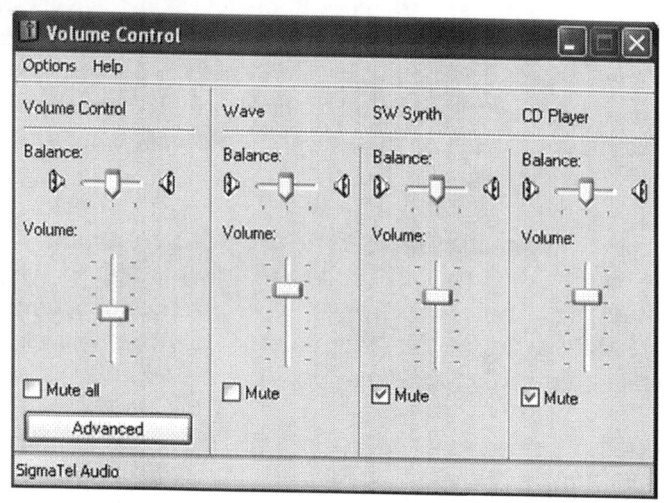

3. Make sure **Volume Control** and **Wave** are not muted. If desired you can mute other audio sources to prevent any bleed over from them. Initially set **Wave** to about 2/3 of its maximum setting.

4. Now we can set the transmitted audio level using EchoLink's Test Server greeting. To verify the drive level you are setting, you will need to monitor the output of your link transceiver with a second transceiver.

Note: Actually the best way to set this is by using a deviation meter or service monitor. If you have one of these pieces of equipment and know how to set deviation you can disregard using a second transceiver as described below.

With a second transceiver tuned to the output of your link transceiver, connect to EchoLink's Test Server. To complete this procedure you will most likely have to connect and disconnect multiple times to generate audio for making the adjustment.

While monitoring the Test Server's greeting, using the *ULI's* **MIC GAIN** control increase the volume to the point where it no longer increases as the control is turned further. You have now reached the point of maximum deviation and some audio distortion may be noted. Now reduce the **MIC GAIN** setting to the point where the Test Server's greeting begins to decrease in volume and then decrease it just a little bit more. The link transceiver's transmit drive level should now be set.

Congratulations! You now have your *ULI* configured for EchoLink operation. If you have not yet already done so, make sure that all the settings found under the various tabs of the Sysop Setup window are properly set.

Now that the *ULI* is configured and receive and transmit levels have been set, you should verify the VOX threshold settings. Setting EchoLink's VOX threshold slider for proper detection of voice signals from your link transceiver, while eliminating other signals is especially important.

If setting up a Repeater Link, pay particular attention to the **Anti-Thump** and **Squelch Crash Anti-Trip** settings found on the **RX Ctrl** tab of the **Sysop Setup** window. Information for making these adjustments is found in the "RX Ctrl Tab" section of Chapter 9.

Chapter 7, Section 3: **SignaLink USB Interface**

TigerTronics has been providing sound card interfaces for a number of years and the *SignaLink USB* is their most recent offering. This product differentiates itself from other interfaces in this chapter because it eliminates using the computer's sound card. It does this by including its own audio CODEC, the equivalent of a "sound-card chip," in its interface hardware, thus divorcing the interface from the PC's sound card.

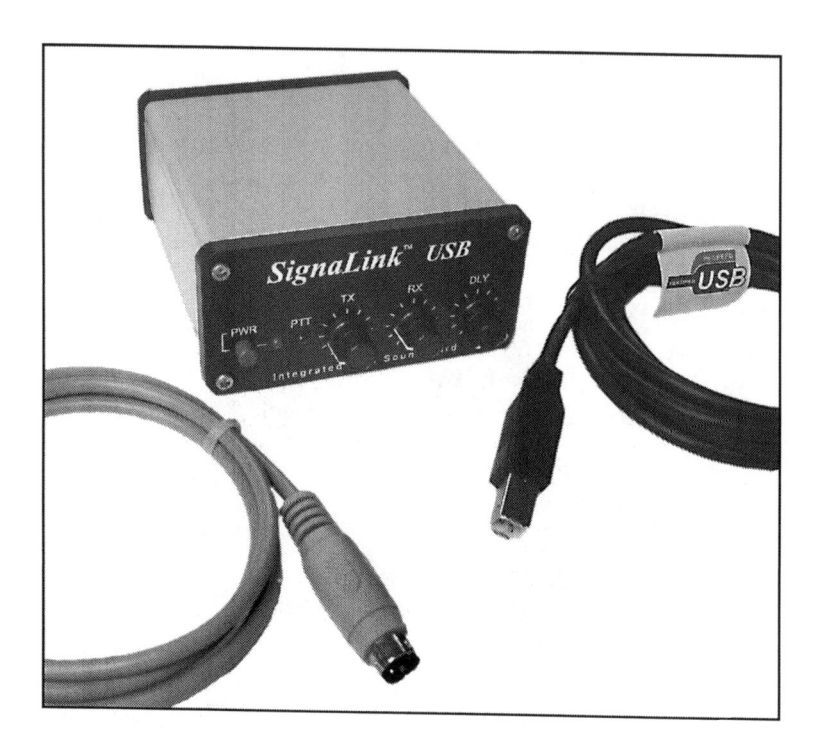

Distinctive Features of the SignaLink USB Interface

The *SignaLink USB* is designed to resolve a number of problems that come up when interfacing a PC's sound card to an amateur radio transceiver. While other products in this section also address these problems, the TigerTronics *SignaLink USB* is a bit unique in how it goes about it.

The following list highlights some of the problem solving features that have been incorporated into the *SignaLink*:

- Incorporating a separate CODEC as a replacement for the computer's sound card prevents interference that can be caused by other applications running on the PC system.

- Incorporating a separate CODEC with its own sound level adjustment settings prevents other PC applications from inadvertently altering the RX and TX sound level adjustments.

- Incorporating a USB chip eliminates the problem of not having serial ports on modern PCs.

- Incorporating a USB chip for digital communication with the PC prevents ground loops from causing audio signal distortion.

- Incorporating a USB interface for communication with the PC allows the USB cable to power the interface, eliminating a "wall wart".

- Incorporating internal VOX capability eliminates the requirement for PC serial port keying.

- Incorporating support for multiple radios via a selection of pre-made interface cables and jumper configurable signals within the SignaLink simplifies interfacing to a radio.

The *SignaLink USB* has one drawback worth mentioning. Since it depends upon VOX as the method of keying, the first syllable of an announcement or conversation may occasionally be clipped.

Despite the above mentioned drawback, the *SignaLink USB* is an alternative worth considering if your computer does not have a serial port or if you don't have access to the COS (carrier sense signal) on your radio. At the time this guide was written, the list price for most models of the *The SignaLink USB* on the TigerTronics web site was $99.95, including your choice of pre-made radio interface cables. In case you need additional radio interface cables, most of them were shown as $14.95 each. More information is available on their web site:
http://www.tigertronics.com/

SignaLink USB Functional Block Description

From a functional standpoint, the *SignaLink USB* is interconnected between the PC and your transceiver as shown in the following figure. The connection to the PC is via a USB cable, which also supplies power to the *SignaLink's* circuitry. Connection to your transceiver's auxiliary data jack or microphone input is via your choice of pre-made cables available from TigerTronics.

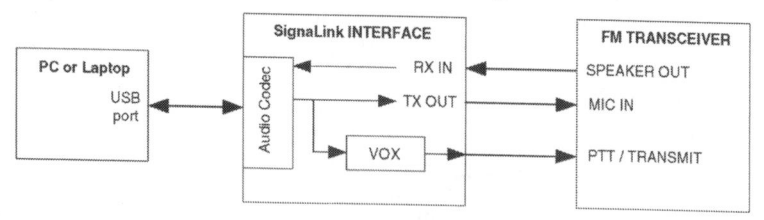

```
                    SignaLink INTERFACE              FM TRANSCEIVER
  PC or Laptop    ┌─────────────────────┐          ┌──────────────┐
                  │           ◄── RX IN ◄───────── SPEAKER OUT    │
       USB   ◄───►│  Audio              │          │              │
       port       │  Codec   ──► TX OUT ─────────► MIC IN         │
                  │                     │          │              │
                  │           ──► VOX   ─────────► PTT / TRANSMIT │
                  └─────────────────────┘          └──────────────┘
```

**SignaLink PC to Transceiver
Interconnect Diagram**

Note that there are no audio cables connected to the PC's sound card. Instead, the TX and RX audio is transported digitally to the *SignaLink* via a USB cable. The audio CODEC in the *SignaLink* converts the digital transmit audio back to analog. Likewise, received audio from the link transceiver is converted into digital for transport to the PC over the USB cable.

Pre-made cables are available for radios that have a data or accessory ports that use a 5-pin DIN, 8-pin DIN, 13-pin DIN, or 6-pin mini-DIN connector. Pin-out variations within a connector type are handled via a configurable jumper block within the *SignaLink*. This allows a single cable to be used with multiple transceivers, even though their connector pin-out configurations may be different.

If connecting to a radio's TNC auxiliary packet radio connector, you probably should use the 9600-baud output in preference to the 1200-baud output. On some radios audio output on the 1200-baud pin has the higher audio frequencies filtered. Great for use with 1200-baud packet operation, but potentially producing muffled sounding voice audio. The radio's volume control does not affect the signal level on either of these outputs.

If your transceiver does not have an auxiliary rear-panel connector available, or if squelched audio is not available from the connector, you most likely will want to use a *SignaLink* cable designed for connecting to the microphone jack. These are available for all radios that use 4-pin round, 8-pin round, RJ11, or RJ45 microphone connectors. If using one of these, an additional audio cable (normally included) is used for connecting to the transceiver's speaker output.

Important: A problem with connecting to an amateur radio's data or accessory port is that they generally output un-squelched audio. Yet, these auxiliary connectors almost always also include a squelch (COS) signal, which could be connected to a PC via its serial port. But using the *SignaLink USB* requires you to develop COS level shifting circuitry and cabling of your own. For the EchoLink software to be able to know when a transmission is being received by your radio, audio going to the PC either needs to be squelched, or a COS / signal connected to the PC's serial port.

If squelched receive audio is not available from your radio's auxiliary port, and you don't want to build a COS cable and level shifting circuitry, you should plan on connecting to the radio's microphone input jack and speaker output.

When ordering a *SignaLink USB* module, it will come with your choice of interface cables. Make sure you select the correct cable for your application.

In the preceding diagram, notice that the transceiver's PTT / Transmit keying is done via a VOX circuit internal to the *SignaLink*. The VOX activates transmission anytime the audio signal from the PC is of sufficient amplitude, thus eliminating a cable to the PC's serial port. Note: this feature is not using your transceiver's VOX, which should be turned off.

On the other hand, receive audio signal detection in the *SignaLink USB* is not dependent upon any level control settings on your PC. The level is controlled via the **RX** knob on the *SignaLink*'s front-panel, and if connected to the radio's speaker output, also by the radio's volume control.

On the transmit side, after presetting your PC's volume control to a suitable gain setting, all future drive level adjustments can be made using the *SignaLink's* front-panel **TX** knob. Note that the *SignaLink*, since it appears to the PC as a second sound card for the computer system, has its own set of PC volume settings.

The front-panel also comes equipped with a transmit indicator LED and a **DLY** knob for setting the VOX hold time, similar to the VOX delay adjustment on an amateur transceiver. Adjustment of the *SignaLink's* **RX**, **TX** and **DLY** knob settings are covered later in this Chapter.

SignaLink USB Installation and Set Up

The *SignaLink USB* comes with well-written and comprehensive *Installation & Operation* instructions. We won't repeat them here, but will briefly outline the procedure and provide information that may help avoid problems when using the *SignaLink* for EchoLink applications.

Installing the Jumpers

The first thing that one must do is open the case and install jumper wires into the socket used to route various signals to the transceiver interface connector pins. This is actually quite simple to do, with TigerTronics providing the required jumper configuration for just about any radio that you might have. If you don't want to configure the jumper header yourself, TigerTronics also provides a number of pre-made jumper headers that can be installed. On the chance that your rig is not covered, they have a section in the manual that helps you determine how the jumpers need to be configured for any radio that you might have.

You will also want to determine if in your configuration you will need to change the settings of any of the three "Special", 2 pin header jumpers that are also on the board:
- **JP2** Increases the receive audio into the *SignaLink's* CODEC
- **JP3** Increases *SignaLink's* transmit audio signal
- **JP4** Increases sensitivity of *SignaLink's* VOX circuit

Installing the Window's Audio Codec Driver

After configuring the jumpers, you connect the *SignaLink* to your computer so that it can automatically detect, locate and install the required drivers for supporting its USB Audio CODEC. Most Windows systems already have the driver; it just needs to be enabled. For Windows 98 installation, the manual says you need the original Windows 98 Installation CD, yet my older Compaq Laptop did not come with one, which gave me some concern. I decided to proceed with the installation anyway, and Windows 98 found the necessary drivers on my system without having to insert an Installation CD.

Sound Card Selection

Once the *SignaLink USB* has been detected and the drivers installed, your system thinks it has two sound cards installed. Having two separate sound cards, one for Windows and another dedicated to EchoLink operation, neatly solves the problem of ensuring that Windows system sounds don't inadvertently get transmitted to the EchoLink network.

This brings us to the next point of the set up process; you need to make sure that for normal Windows system and application purposes, that the original sound card that came with your computer is selected. Likewise, for purposes of EchoLink, that the USB connected audio CODEC internal to the *SignaLink* is selected. If the system's sound card is not properly selected, your system will be unable to produce any sounds—a sure sign that you need to configure the setting.

For the following steps, make sure that the *SignaLink USB* is plugged into your system's USB port.

The sound card selection instructions that come with the *SignaLink* can be followed in most instances. It's a matter of going to the **Control Panel** and configuring the default **Sound Playback** and **Sound Recording** devices found under the **Audio** tab of the **Sounds and Audio Devices** properties set up window. If you have Windows 98, though, instead you will use the same method to access the **MultiMedia Properties** window.

When you have located the correct set up window, use the pull-down selection box to set both the **Recording** and **Playback** devices to the sound card that came with your computer. The list should be short, probably with only two or three entries. The name for the system's sound card will vary, on my Windows 98 system it was called **Via Audio (WAVE),** on my Windows XP system it was called **SigmaTel Audio.**

> **Note: If using Vista,** right click on the speaker icon in the task bar and select **Sounds,** then the **Playback** tab. Select your computer's sound card from the list and set it back to the default sound card for playback. Next select the **Recording** tab, and select your computer's microphone as the default card for recording.

Windows 7 instructions can be found on the TigerTronics web page.

Once you have accomplished the sound card selection set up, your computer is configured to send all system sounds to your computer's sound card and speakers. Just like it was before installing the USB Audio Codec drivers.

Now we still have to tell EchoLink which sound card to use. Click on the **Tools** button and select **Setup** from the pull-down menu. On the **System Setup** window that opens, select the **Audio** tab and make the following selections:
- **Input Device**   USB Audio Device (or some similar name)
- **Output Device**   USB Audio Device

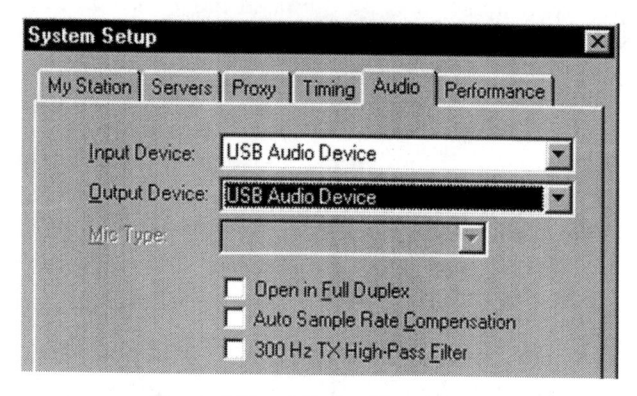

Sound Card Selection

Sysop Settings for the SignaLink USB

A few Sysop settings must be configured for the EchoLink software to work properly with the *SignaLink USB* interface.

Under the **Tools** menu, select **Sysop Settings**, and on the window that opens, select the **RX Ctrl** tab. Here select **VOX** as the received **Carrier Detect** mechanism. This instructs EchoLink to use its internal software based VOX mechanism to detect when a valid audio signal is being received from the link transceiver. A little later in this section, you will also need to adjust the VOX level slider found under the audio level meter. Additionally, as initial presets set the remainder of the fields as shown in the figure below.

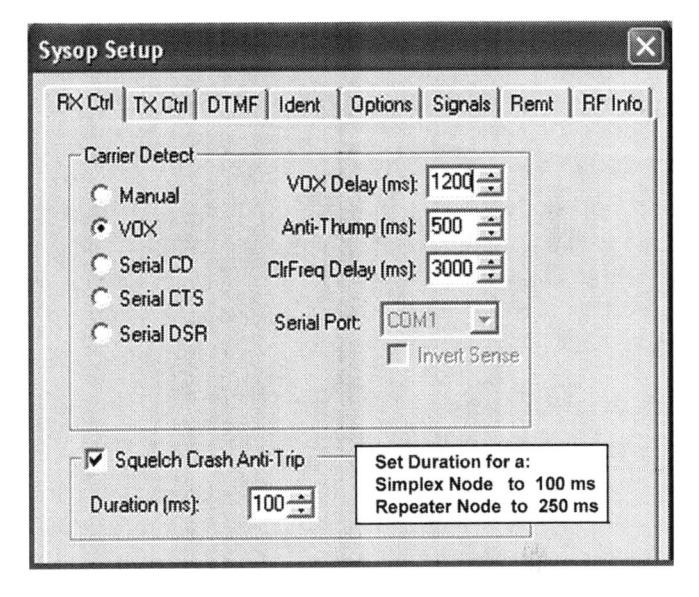

Receive Control settings for use with SignaLink

The **Squelch Crash Anti-Trip** setting prevents unwanted noises from being transmitted to the Internet. For simplex node operation, the **Duration** can be set to 100 ms, for repeater nodes set it to 250 ms.

Next select the **TX Ctrl** tab, and select **External VOX** as the **PTT Activation** method. This informs EchoLink that the *SignaLink* will be using its own VOX mechanism to trigger transmission of your link transceiver when audio is received from the Internet.

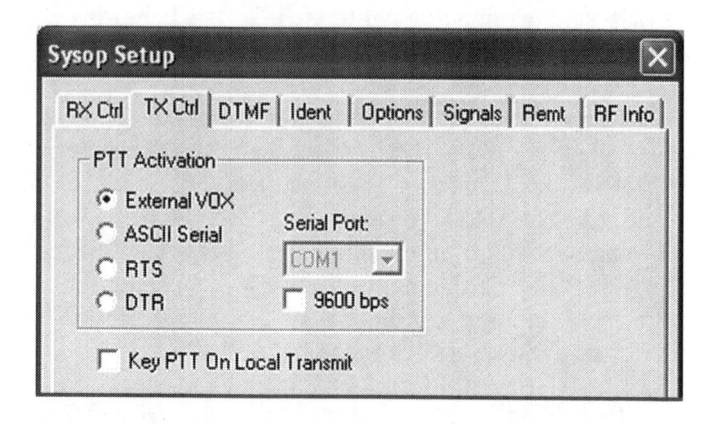

Sysop Setup

RX Ctrl | TX Ctrl | DTMF | Ident | Options | Signals | Remt | RF Info

PTT Activation
- (•) External VOX
- ( ) ASCII Serial
- ( ) RTS
- ( ) DTR

Serial Port: COM1
[ ] 9600 bps

[ ] Key PTT On Local Transmit

Transmit Control settings for use with SignaLink

Setting Received Audio Drive Level from the Transceiver

Adjustment of the received audio level from the Link Transceiver is performed using a combination of the *SignaLink's* front-panel **RX** level control knob, and if connected to the radio's speaker output, the radio's Volume Level Control. The PC's recording level control is not used and does not need to be set.

RX Drive Level Procedure:

1. Preset the *SignaLink's* **RX** control knob to mid range.

2a. If you are setting up your link transceiver as an EchoLink simplex node, using another radio transmit on the link's simplex frequency.

2b. If you are setting up your link transceiver as a repeater link node, make the following adjustment while "normal" level audio is being transmitted from the repeater.

3. While the link transceiver is receiving a signal, observe EchoLink's audio level meter. If the *SignaLink* is connected to the link transceiver's speaker output, adjust the radio's volume control so that audio peaks are starting to overdrive the meter (a yellow bar indicates overdrive), then back off until audio peaks only overdrive the meter on loud audio peaks. If not connected

to the transceiver's speaker output, use the *SignaLink's* **RX** control to accomplish the same objective. If you don't have enough range, you might need to install **JP2**.

Setting EchoLink's Received Audio VOX Threshold

Once the receive drive level has been set, we can set EchoLink's received signal VOX threshold level. It's important to set the VOX threshold level sufficiently high to prevent undesired EchoLink transmissions due to receiver noise pickup. On the other hand it needs to be sensitive enough to reliably trigger transmission when receiving weak signals from the radio.

When in Sysop mode, an adjustment slider is displayed under the audio-level meter. When the audio level exceeds the VOX threshold level set by the slider, EchoLink will transport audio being received from the link transceiver to the Internet. It's not necessary to be in an actual QSO with a remote station to perform these adjustments; monitoring local link transceiver activity is sufficient

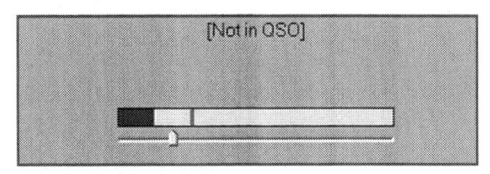

Setting Simplex Node VOX level:
Setting the VOX trigger level for Simplex node operation is rather simple, just set it about halfway between an un-modulated carrier level and the level seen when receiving a fully modulated carrier. While operating, you can increase the sensitivity if transmission is not being triggered reliably on weak signals, or decrease the sensitivity if experiencing false triggering.

Setting Repeater Node VOX level:
Setting the VOX trigger level is a bit more critical for repeater node operation. While monitoring your repeater, make note of the difference in signal levels when receiving voice, repeater IDs and courtesy tones. Ideally repeater ID and courtesy tones will produce lower level signals than voice.

If there is a pronounced difference between voice and the repeater's IDs and courtesy tones, it should be possible to set the VOX threshold between these levels and prevent unwanted signals from being transmitted over the EchoLink network. This will not, however, eliminate ID's that occur concurrently while a user is transmitting.

If there is no significant difference in the levels, perhaps the repeater controller can be adjusted to reduce the volumes of the repeater ID and courtesy tone. Failing that, you will need to consider alternative methods for keying transmission to the EchoLink network. Refer to Chapter 10 for a discussion of using CTCSS or repeater co-location as methods of solving the problem.

For repeater operation, make sure the **Squelch Crash Anti-Trip** setting on the Sysop **Rx Ctrl** tab is initially set to 250 ms, which is a typical setting. If the VOX trigger level has been set correctly and you are still having trouble with courtesy tones or squelch crashes triggering transmission, this delay can be increased.

For detailed instructions on setting the VOX level, refer to the **Repeater Linking Tips** section of the EchoLink help files.

When you think you have the VOX set properly, verify that when you connect to the EchoLink Test Server, that after hearing the welcome message that the repeater stops transmitting and remains idle. Recording and playing back your own voice should also operate normally and not result in unnecessary additional transmissions from either the Test Server or your station.

Setting SignaLink's VOX and the Transmit Drive Level

Adjustment of the transmit audio level going to the link transceiver is performed using a combination of the *SignaLink's* front-panel **TX** level control and the computer system's output volume level control. The computer system's volume level control is used as a master overall range setting and for reliable triggering of the *SignaLink's* VOX circuitry (VOX gain). Once set, you can generally forget about it and use *SignaLink's* **TX** drive level control to make any needed transmitted audio level adjustments. The computer system's volume level setting affects the *SignaLink's* internal VOX level / gain setting, so it needs to be adjusted first.

VOX Adjustment and TX Drive Level Procedure:
1. Preset the *SignaLink's* **TX** control knob to mid range.

2. Preset the *SignaLink's* **DLY** control knob to about 1:00. This control sets the VOX dropout delay between words. The delay needs to be long enough to keep the link transceiver transmitting during normal speech, only dropping during longer pauses in transmission. The 1:00 position seems to be a good starting point. Later, after the system's volume level has been set, the VOX delay will be fine-tuned in step 7 of this procedure.

3. Now verify that the *SignaLink* is selected as the sound card device by selecting **Adjust Sound Device** found under the **Tools** pull-down menu. Select the **Playback Control** option, which brings up the **Speaker** volume adjustment window. Next click on **Options** to bring up the **Properties** window, as shown in the following figure. This may appear slightly different on different versions of Windows.

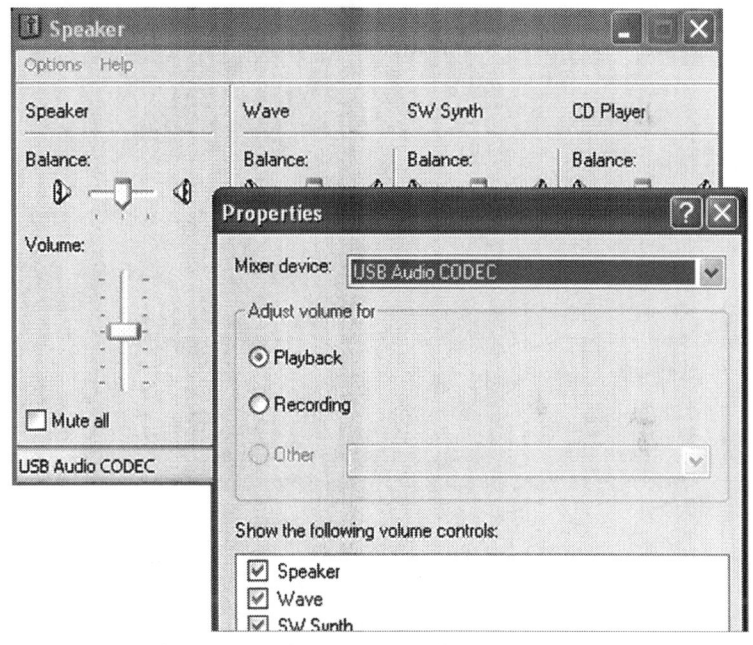

SignaLink Sound Card Selection

Make sure that **USB Audio CODEC** (the *SignaLink* device) is selected as the **Playback Mixer device** and then press **OK**.

Page 101

4. Leave the **Speaker** volume control window open. Next we will use EchoLink's **Spoken Voice** station ID as a test signal. Under **Tools**, select the **Ident** tab of the **Sysop Setup** window. Select **Spoken Voice**.

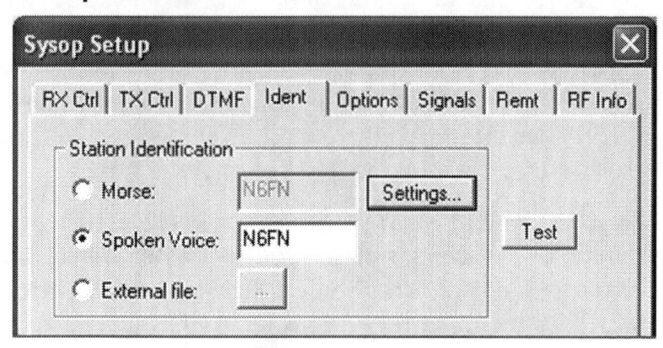

Using Spoken Station ID as a Test Signal

5. While repeatedly pressing the Test button to verbally transmit your call sign, adjust the **Speaker / Wave** volume control so that the *SignaLink's* VOX is just being reliably triggered, (the red PTT LED lights). Then set the **Speaker / Wave** volume level control above that level by one or two tick marks of the **Volume** level scale.

6. Now we can set the transmitted audio level using EchoLink's Test Server greeting. To verify the drive level you are setting, you will need to monitor the output of your link transceiver with a second transceiver.

Note: Actually the best way to set this is by using a deviation meter or service monitor. If you have one of these pieces of equipment and know how to set deviation you can disregard using a second transceiver as described below.

With a second transceiver tuned to the output of your link transceiver, connect to EchoLink's Test Server. To complete this procedure you will most likely have to connect and disconnect multiple times to generate audio for making the adjustment.

While monitoring the Test Server's greeting, using the *SignaLink's* **TX** knob increase the volume to the point where it no longer increases as the control is turned further. You have now reached the point of maximum deviation and some audio distortion may be noted. Now reduce the **TX** knob setting to the point where the Test Server's greeting begins to decrease in volume and then decrease it just a little bit more. The link transceiver's transmit drive level should now be set.

7. With the TX audio levels set as described above, you can now verify the VOX **DLY** setting. Experiment using a slow count (increasing the spacing between numbers) and readjust the **DLY** knob as required for proper VOX operation. Either increasing the delay if you find transmission dropping out while you are still speaking or decreasing the delay if transmission seems to be holding too long after a pause in speech.

Congratulations! You now have your *SignaLink* configured for EchoLink operation. If you have not yet already done so, make sure that all the settings found under the various tabs of the Sysop Setup window are properly set.

Now that the *SignaLink* is configured and receive and transmit levels have been set, you should verify the VOX threshold settings. Setting EchoLink's VOX threshold slider for proper detection of voice signals from your link transceiver, while eliminating other signals is especially important.

If setting up a Repeater Link, pay particular attention to the **Anti-Thump** and **Squelch Crash Anti-Trip** settings found on the **RX Ctrl** tab of the **Sysop Setup** window. Information for making these adjustments is found in the "RX Ctrl Tab" section of Chapter 9.

Chapter 7, Section 4: Other PC to Radio Interfaces

The manufacturers mentioned below provide a variety of products, ranging from relatively simple interfaces to rather complex interfaces that attempt to do everything.

RIGblasters by West Mountain Radio
http://www.westmountainradio.com/
West Mountain Radio has several models in their very popular *RIGblaster* series of interfaces, which are USB connected. *RIGblasters* can be quite sophisticated, like the *RIGblaster Pro* or the *RIGblaster Duo,* which have a very broad and interesting set of features. They also have more basic models, like the *RIGblaster Nomic* and *RIGblaster Plug & Play,* which are lower cost, barebones simple interfaces to operate. Prices range from about $70 to $350. These products are available from most ham radio equipment distributors.

Sound Card to Rig Interfaces by MFJ
http://www.mfjenterprises.com
MFJ manufactures several different rig interfaces, priced from about $60 to $150, depending upon the features supported. These products are available from many ham radio equipment distributors.

Special Purpose EchoLink Interfaces by KJ6ZD
http://www.kj6zd.net
Norbert, KJ6ZD, manufactures a line of repeater controllers and EchoLink interfaces under the *Amateur Radio Accessories* brand name. He has made several refinements since the first EchoLink interface he came out with in 2004. At the time this is being written, the currently available interface is the EI-151, available for $99.95. To be released late in 2010 is his latest creation, the EI-160. Both of these products are special purpose interfaces designed specifically for EchoLink operation.

The EI-160 is interesting because it is relatively easy to set up, has all the necessary features for EchoLink operation, including COR / COS support and a unique ALC (automatic audio level control) on the RF transmit side and on the receive side for hardware DTMF tone detection. Also helpful is a programmable transmit time-out timer, which shuts down the link transmitter without disconnecting the EchoLink user.

Because the user manuals for EI-151 and EI-160 products are geared specifically for EchoLink operation, they are quite explicit as to what you need to do to set up either simplex or repeater nodes.

Special Purpose EchoLink Interface by VA3TO
http://www.ilinkca.com

Like the WB2REM ULI interface board described earlier in this chapter, the VA3TO board has been around for a number of years and is also mentioned in the EchoLink help files. This board is available as a high quality kit. In addition to EchoLink, this board can also be used with any of the other digital modes of operation. The board features COR / COS connect capability, a four minute maximum transmit timer, on-board DTMF decode chip and a serial interface. Completely tested and in an enclosure it's priced at $85 plus shipping. Kit versions vary in pricing depending upon how complete the kit is.

Special Purpose EchoLink Interfaces by WB2REM / G4CDY
http://www.ilinkboards.com

Jim, WB2REM and Terry, G4CDY have partnered to manufacture a line of EchoLink interfaces. Their ULI interface has long been a favorite of those wishing to use a special purpose interface, and is discussed in detail in Chapter 7, Section 2. As this is being written they are in the process of developing a new and improved interface, expected to be in production sometime in 2011. While the new interface retains all the capabilities of the original ULI, the new device includes the following additional features:

- USB port, eliminating the need for a serial port to USB adaptor.
- USB powered, eliminating need for a separate power connection.
- Built-in, low-noise sound card providing improved audio quality.
- Built-in sound card eliminates need for cables to the PC's sound card.
- Plus several other miscellaneous additional improvements.

Chapter 8: **Link Radio Requirements**

A variety of different radios can be used as simplex node or repeater link transceivers. In addition to amateur transceivers, surplus commercial transceivers can be used. The advantage of using decommissioned commercial radios is that some of them can be obtained quite cheaply on the Internet, frequently for under $40. Often commercial radios are more capable of handling extended periods of transmission and frequently provide the much desired COS signal (received carrier sense signal) at one of their external connectors.

If you care to spend the money, some dual-band amateur transceivers can be ideal for use as link radios when implementing an auxiliary control link for remote control operation. While one band is used as the link frequency the other band can be monitoring for DTMF tones on an auxiliary control frequency. Since dual-banders often combine their audio signals into a composite signal, they can eliminate the need for an external audio mixer. Of course, the alternative to using a dual-bander is to use two single-band radios, which can be inexpensive when making use of surplus commercial radios.

Output Power Rating

When nodes are connected to conference servers they can be transmitting for long periods as different conference participants take turns transmitting. As a consequence, selecting a transceiver that can withstand high transmit duty cycles is important. To help solve this problem, you can run at reduced power levels, perhaps taking a 50-watt rated transceiver and running it at five watts. Adding an external fan on the heatsink is another method of helping the radio survive high transmit duty cycles.

CTCSS, Tone Squelch

Using CTCSS (tone squelch, sometimes also called PL) as a method of eliminating unwanted signals is highly desirable. CTCSS encode and decode is available on most all VHF/UHF amateur transceivers and also frequently found on surplus commercial gear.

Carrier Operated Switch, COS

Especially when setting up a repeater link, it may be desirable to have a COS (carrier operated switch) signal available at one of its auxiliary connectors. This signal is used to tell an external controller that the squelch is open and a signal is being received. Typically the COS signal is wired to the PC's serial port so that the EchoLink software can reliably tell when a signal is being received. Most commercial transceivers and some amateur transceivers have this capability. You will need to look at the documentation carefully, though; this signal often goes by a variety of names: squelch, COR, COS, CAS, busy and others. If not available at an external connector, an alternative is finding a place in the circuitry where you can tap into an equivalent signal, perhaps the busy light or some other point.

If you intend to use CTCSS, ideally you want a radio that produces a COS signal that only goes active when the signal being received has a matching CTCSS tone. Most amateur radios only sense that a signal is being received, and do not include CTCSS in the COS signal. It is more common to find the COS signal being conditioned with CTCSS in commercial radios.

A potential problem in using COS from a link transceiver is if the repeater has a long squelch tail (hang time). As long as the COS signal is active, EchoLink will keep transmitting to the Internet. So if the squelch tail is any longer than a second or two, other stations will be prevented from transmitting until some time after the COS signal finally drops.

If the repeater tail is too long, perhaps it can be shortened via the repeater controller's configuration settings. If not, then you may need to use EchoLink's **Smart VOX** capability instead of relying on the COS signal. If configured properly, EchoLink should be able to ignore a repeater's squelch tail, squelch crash and courtesy tone.

Alternatively, if your system will be using CTCSS, and you are connecting to the receiver's auxiliary speaker output connector, COS may not be required. The speaker audio will already have the equivalent of COS and CTCSS applied to the audio signal. Though, when using speaker output without the benefit of a COS signal, the EchoLink software needs to be set for VOX operation, which may not be 100% reliable and adds some delay to the turnaround time between reception and transmission. These delays can be shortened by a second or more when COS is available.

There are a couple of drawbacks to using EchoLink's built in **Smart VOX** capability. One is that the **VOX Delay** adds a second or two after each transmission. Excessive delays adversely affect the flow of conversation, requiring longer pauses on the part of all participants. As the delays tend to add up in the overall EchoLink system, they should be minimized whenever possible. Another difficulty is that EchoLink's internal VOX depends upon the received signal's audio modulation level (amplitude). The COS signal may be more reliable, since it tracks the receiver's squelch. If connecting to the COS signal is not practical with the equipment you have, go with EchoLink's **Smart VOX**.

Received Audio Characteristics

When selecting a link transceiver, some consideration should be given to the characteristics of the transceiver's received audio, and where it is picked off. If the receive audio is taken from an auxiliary speaker-out jack, it will be quieted (cut off) whenever received signals are below the squelch level, and if CTCSS is being used, whenever a matching tone is not being received. Which is what you want for EchoLink operation.

Attributes of the audio quality taken from an auxiliary rear-panel jack, generally intended for packet operation, may vary from radio-to-radio. It is common for audio taken from a rear-panel jack to have one or more of the following limitations:

- Audio un-squelched, producing "white noise" when a signal is not present.

- Audio un-conditioned with CTCSS. Even though the repeater is transmitting a CTCSS tone, and your transceiver is programmed to use it, frequently it's not used to gate the audio going to the auxiliary jack connectors.

- Audio incapable of being adjusted. Front panel volume controls generally have no effect on signals obtained from the rear panel jacks.

- Audio output might be bandpass filtered, fine for packet but imparting a muffled audio quality.

While amateur radios often suffer from one or more of the above problems, some of these issues may not be important depending upon the transceiver-to-PC interface hardware you are using. In any case, if these limitations are a problem, taking the audio from an external speaker jack generally resolves the issue.

Power Supply Requirements

When setting up a node, don't overlook the power supply requirements. You will want one that is reliable and that won't overheat during long periods of transmission. It's also vital that if the power mains (AC input voltage) go off for any reason that both the radio and the power supply come back up in the on state. Some radios may not power back up, or even worse, come back up on the wrong frequency. Alternatively, you may want to consider using an uninterruptible power supply, which makes use of a battery when the power is off. As a part of setting up your node, you will want to verify what happens to the node when power is lost.

Amateur Transceivers

Many amateur transceivers can be successfully used as link transceivers. You may already have an old radio you want to use. If not, used ones can be bought on eBay or from classified ads posted on www.eHam.com or www.QTH.com. Review the above requirements when making a selection.

Of note, the Kenwood TM-V71A and TM-D710A dual-band mobile transceivers have capabilities especially for supporting EchoLink operation. To simplify node access while operating mobile, the radios provide 10 special memory locations designed to store EchoLink node numbers. Once programmed, you can easily have the radio transmit the DTMF sequences for connecting to remote nodes. While the DTMF capabilities are useful for mobile operation, these radios also have a built-in transceiver-to-PC interface useful for setting up your own node. Using the optional PG-5H cable provided by Kenwood the radio can be directly connected to a PC's sound card and serial port, simplifying the creation of your own simplex or repeater nodes. Being dual-banders, the same radio can also be set up for auxiliary control of the node.

Auxiliary Data Connectors on Amateur Transceivers

Most amateur transceivers come equipped with a TNC compatible auxiliary data / packet connector. The signals on this connector are relatively well standardized, typically using a 6-pin or 8-pin mini-DIN connector at the rear of the radio. Instead of connecting to the speaker-out and microphone jacks, many node builders have used this connector for connecting their link transceiver to the node's EchoLink interface circuitry. Usually, PTT and squelch out (the equivalent of a COS, carrier sense signal) are also available on this connector.

When considering using this connector, it's important to remember that the audio-out signal is almost always un-squelched. Neither is it conditioned by the presence of a matching CTCSS tone. Audio should only be sent to EchoLink while a valid signal is being received. If the audio is not suppressed when a signal is not present, "white" noise will prevent EchoLink's VOX from operating properly. By developing some circuitry, this problem can be overcome by using the squelch out signal to gate the received audio signal.

Be aware that if connecting to an auxiliary TNC style connector, that generally you should set your radio for 9600-baud operation in preference to 1200-baud operation. Audio output on the 1200-baud output pin could have the higher audio frequencies filtered which can lead to a muffled sounding voice audio.

An additional point of consideration is the audio levels at the radio's connector. The audio output level is generally fixed, and cannot be adjusted at the radio. When a radio is set for 9600 baud operation, the 9600 baud output pin generally provides about a 500 mV peak-to-peak receive signal. The 1200 baud output receive signal is typically about 300 mv pp. Transmit input signal requirements may vary by radio type, and can range from 0.4 Vpp to 2 Vpp. Depending upon the radio, the above described signal levels may be different. You may want to verify that the interface you select can operate properly with the signal level provided by your particular transceiver.

Surplus Commercial Transceivers

Many surplus VHF/UHF mobile transceivers can be readily adapted for use as VoIP linking transceivers. Indeed, there are many amateur radio repeater builders and VoIP node owners who are adamant about the benefits of using surplus GE and Motorola transceivers, not the least of which are easy adaptability, low cost and high reliability.

These transceivers can often be found for under $40 on eBay or from classified ads posted on **www.eHam.com** or **www.QTH.com**. As shown in the URL's listed below, there is a great deal of information available on the Internet explaining how to configure these radios for amateur radio use.

One disadvantage of using these radios is that they require PC software for programming their frequency channels. Unlike amateur radio transceivers, commercial radios are not designed to be programmed from the front panel. For some models, the software can be found on the Internet, for others you may need to have them programmed by a shop that services commercial radio equipment. Yahoo Groups can be helpful for locating the required software.

Another possible disadvantage is if you are unfamiliar with reading schematics and troubleshooting. These radios may require some repair, circuit modifications and custom cable making. If you are not comfortable with those activities you might want to stick to using a known-working amateur transceiver.

http://www.irlp.net/
Selecting the **Node Radios** link found at the above URL provides information on both amateur radios and surplus commercial gear that has been successfully used for VoIP linking. While the site is dedicated to IRLP, the linking radio requirements are essentially the same for EchoLink.

http://www.kd4raa.net/linkradios/moto-irlp.htm
KD4RAA provides some good information on Motorola M series radios on his web page. He also sells pre-made interface cables, and if you want to make your own, a schematic of how to make the cables.

http://www.qsl.net/w4xe/new_page_3.htm
http://www.powerlinenoise.com/n0rq/MVS/index.htm
The above two links provide a good deal of information for converting GE / Phoenix MVS series of radios to amateur use.

In addition to the above web pages, the **Repeater Builders** web site http://www.repeater-builder.com/rbtip/ provides a treasure trove of information on all kinds of radios, amateur and commercial.

Commercial Transceiver Yahoo Groups

Yahoo Groups is another good source of information for converting surplus transceivers into linking transceivers for VoIP node operation.

The **ge_mvs** group in addition to postings about various problems and solutions has download files available for the GE MVS manuals, schematics and programming software.

The **Repeater-Builder** group is a good source of information for all sorts of repeater related topics.

The **Radio-Programming2** group concentrates on providing two-way radio programming information and help.

The **Motorola-Radius** group has information on using the Maxtrac, Radius, GM300 and similar equipment for base, mobile, link, or repeater stations.

Chapter 9: Sysop Mode Setup

Once you have determined the type of PC-to-transceiver interface and the transceiver you will be using, you should verify the configuration of all Sysop menus. The **Sysop Setup** screens are found under the **Tools** pull down menu.

Information available using the **Help** button on each of the Sysop Setup windows should be adequate for setting these menus, and we will not repeat that information here. A short summary of each of the screens is presented below. Additional commentary is provided, as it seems necessary or helpful.

RX Ctrl Tab

The Sysop receive control screen is used to configure how audio from the link transceiver is detected. Various parameters can be set to ensure that undesired squelch tails, repeater ID's, courtesy tones and noise bursts are prevented from being transported over the EchoLink network.

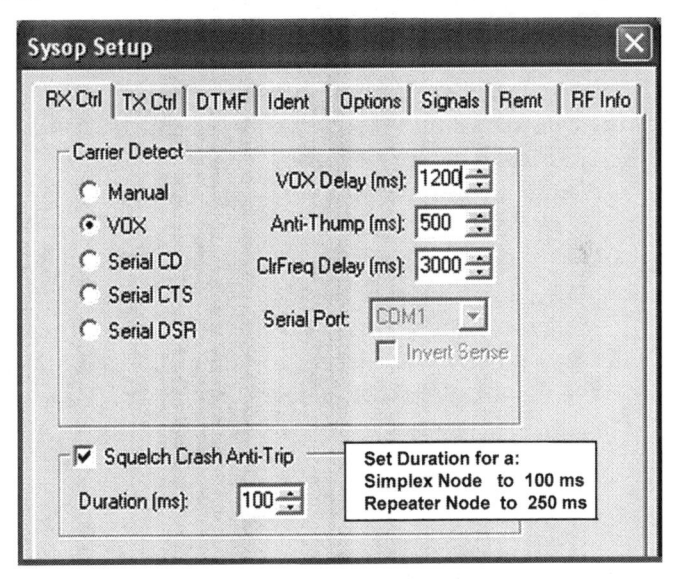

Sysop Receiver Controls Window

When a COS (carrier-detect signal) is not available from the link transceiver the **VOX** selection and its accompanying **VOX Delay** and the VOX trigger level adjustment under the voice-level meter can be used to trigger audio transmission to the Internet.

If a COS signal is available, it can be connected to any of three serial port pins on the PC: CD, CTS or DSR. You most likely will need a level shifting circuit in between the radio and the PC. Some interface boards provide the necessary circuitry. Refer to Appendix A for PC serial connector pin out information.

For detailed instructions for setting the VOX levels, refer to the *"Setting the Received Audio Drive Level"* and *"Setting EchoLink's Received Audio VOX Level"* procedures in Chapter 7. If using a general-purpose PC-to-transceiver interface, refer to the instructions found under the *RIGblaster Pro II* section of Chapter 7. If using either the *ULI* or *SignaLink* interfaces, refer to the instructions found in those sections.

The **Anti-Thump** delay prevents EchoLink's VOX from being triggered by the link transceiver's squelch crash. The default **Anti-Thump** delay of 500 ms is a good starting point, but it may need to be increased. When a repeater has been re-transmitting the signal being received by the link transceiver, the crash is heard on the output of the repeater. If the VOX is triggering on the squelch crash, the **SIG** indicator at the lower right of the EchoLink window flashes when the link stops transmitting. To prevent this, gradually increase the **Anti-Thump** delay until the **SIG** indicator no longer flashes.

When operating as a repeater node, the **Squelch Crash Anti-Trip** setting prevents EchoLink's VOX from being triggered by short courtesy tones and squelch crashes as the repeater drops its carrier. A setting of 100 ms should be adequate for simplex links, and 250 ms is a good starting point for repeater nodes. Try increasing the delay if courtesy tones or squelch crashes appear to be triggering the VOX.

TX Ctrl Tab

The Sysop transmit control screen configures the method EchoLink uses to key the link transceiver. Three methods are provided:
- VOX circuit external to the EchoLink application,
- ASCII serial data commanded over the PC's serial port,
- Direct signaling via one of the PC's serial port handshake lines.

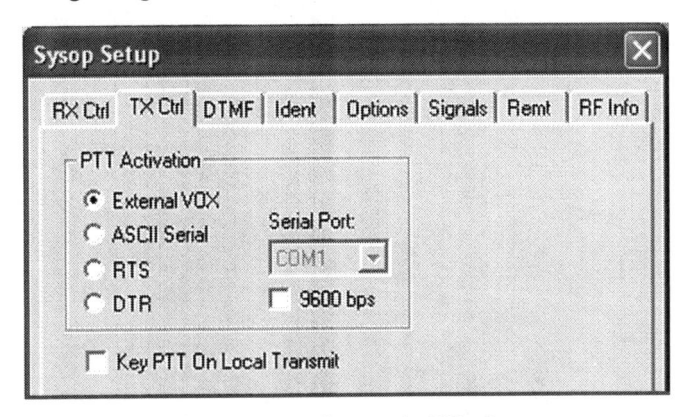

Sysop Transmit Controls Window

Unless you are using the WB2REM, ULI interface board, or some other board with an ASCII serial interface, you will select either **External VOX** or the **RTS**, **DTR** direct signaling method.

DTMF Tab

The **DTMF** tab, among other things, allows you to either accept or modify EchoLink's default commands. Unless there is a reason to modify the commands, such as interference with a repeater's other DTMF services, it's probably best to stay with the default commands that EchoLink users are familiar with. The one exception would be assigning your own codes for bringing your link up (**LinkUp**) and shutting it down, (**LinkDown**). You may not want everyone knowing those, but be aware that since the DTMF control codes are transmitted over the air, that it's possible for someone to figure out what they are.

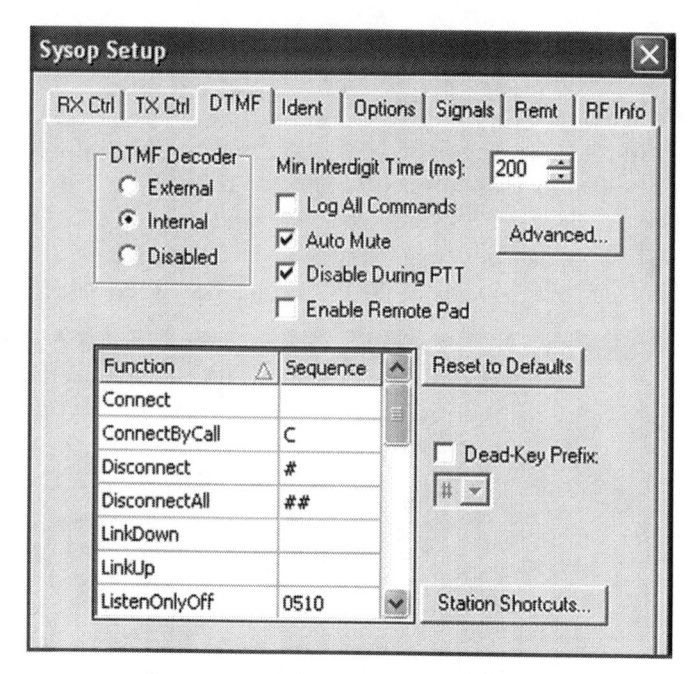

Sysop DTMF Control Codes Setup

If you are having trouble with DTMF digits being interpreted properly, you might want to select the **Advanced** button and try fine tuning the DTMF decode parameters.

EchoLink's software DTMF detection method can be susceptible to incorrect operation with weak or noisy signals. Some special purpose EchoLink hardware interfaces, including WB2REM's, *ULI* interface and KJ6ZD's *EI-151* and *EI-160* interfaces contain hardware DTMF decoders which should solve this problem.

The **Station Shortcuts** button can be used to configure short DTMF command sequences for connecting to EchoLink nodes that are frequently accessed. These might be useful for joining scheduled nets, conference servers or for connecting to a designated node for emergency purposes.

Unless you have a specific reason for broadcasting DTMF tones received on your link transceiver out over the Internet, you should enable **Auto Mute**. If DTMF tones local to your RF node are sent to connected conference servers or repeaters, it may disrupt conversations or inadvertently activate DTMF controls on remote systems. Check **Auto Mute** to prevent this from happening.

Similarly, it's good idea to check **Disable During PTT** to prevent EchoLink from decoding DTMF signals while your link transceiver is transmitting. This inhibits DTMF signals received from the Internet from being decoded and potentially affecting the operation of your node.

As part of setting up your link, you will most likely want to be able to enable and disable your link by using DTMF commands. Control can occur either on the link's receive frequency or via an auxiliary control receiver operating on a control link frequency. (Note: typically an auxiliary control link requires that your radio-to-PC interface have an audio mixer for combining the main and auxiliary audio paths.) Enter separate password codes for the **LinkUp** and **LinkDown** DTMF commands.

If you are setting up a repeater node and are not hearing DTMF codes on the output (necessary for your link transceiver to hear them), your repeater controller is most likely equipped with a feature that mutes them. If this is the case, you will may want to enable the **Dead-Key Prefix** option. Refer to the EchoLink help file for full details on setting this up.

Ident Tab

To comply with the FCC's or foreign country station identification requirements, your station's call sign and the parameters for configuring when it is transmitted are configured using the **Ident** tab. Generally the call sign used will be your call sign, even when the station is being used as a link to a repeater.

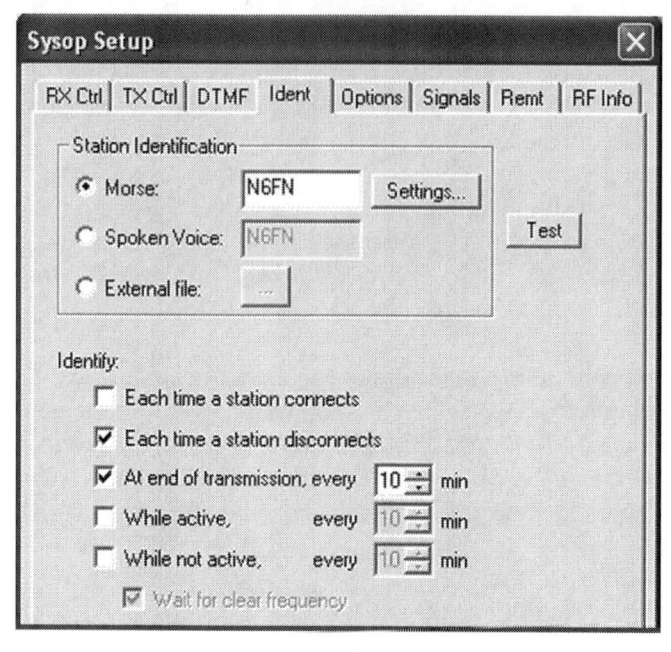

Sysop Station ID Setup Window

You have a choice of three methods for how your station's call sign will be transmitted:
- **Morse Code** (less interference with voice communication)
- **Spoken Voice** (uses EchoLink's internally generated letters)
- **External File** (you can store a custom .wav audio file)

There are also several selections determining when your call sign will be transmitted. Be careful of using **While active** and **While not active**, to avoid other amateurs becoming disgruntled with unnecessary station identification broadcasts on either simplex or repeater frequencies. On the other hand, **While not active** might be handy to use in remote areas as a method of "advertising" your simplex link. In urban areas, where finding a free simplex frequency can be difficult, its probably best not to ID unless a link is actually being made.

When setting up this screen its good to remember that according to FCC regulations your station must identify at least once every ten minutes during a communication and also at the completion of a communication. You should ensure that long-winded users do not prevent the station from identifying within the ten-minute period. This can be accomplished by using the **Options** tab, where a maximum transmit time can be configured.

Options Tab

The **Options** tab has settings for configuring announcements that can be made when stations connect and disconnect from your node. There is also a setting for setting the maximum amount of time that your link transceiver can be keyed. This is similar to the "alligator" timer commonly found on local repeaters. With one important difference, though, in addition to cutting off the transmission, EchoLink also disconnects the station from your node. In light of that, you might want to make this time a little longer than the 3 minutes commonly found on repeaters. You might not want to disconnect someone if he just got a little long winded.

Several of the special purpose EchoLink hardware interfaces, WB2REM's, *ULI* interface, the *VA3TO* board and KJ6ZD's *EI-151* and *EI-160* interfaces contain hardware transmit time-out timers which should solve this problem. When the time-out timer is set to a time shorter than the one EchoLink is set to, transmission from the link transceiver is cut off, without disconnecting the user from your node. If this should occur, other users will have to notify the person that his transmission had been cut off.

As an example: When using one of the boards containing a hardware transmit time-out timer, it could be set to 3 minutes; then the EchoLink timer could either be disabled by entering "0", or set to some maximum time, like 5 or 6 minutes. Anyone talking longer than 5 or 6 minutes probably ought to be disconnected from your node!

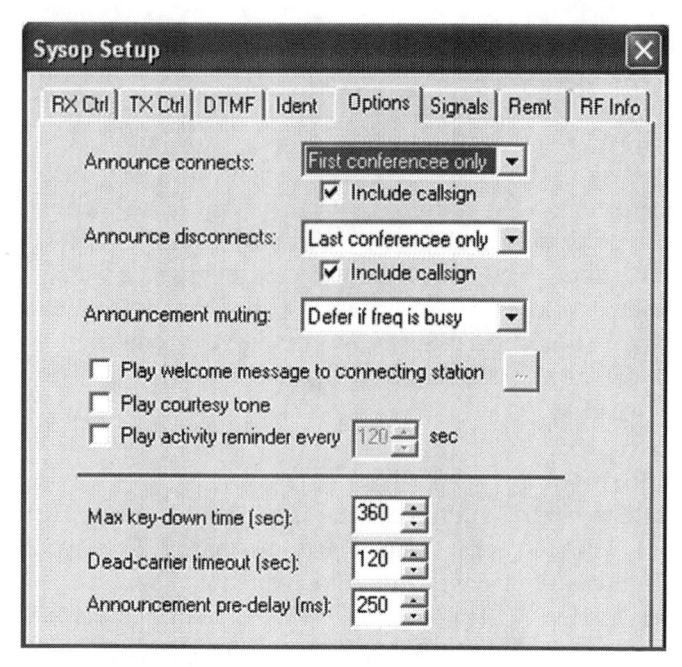

Sysop Connection and Keying Controls Window

How many connects and disconnects you wish to have announced is a personal preference. You may want them all announced, which might adversely affect the flow of a roundtable discussion, or you may wish to limit them. A lot depends upon how your system will normally be used.

Likewise, the use of welcoming messages and courtesy tones is a matter of personal preference, either your own, or the preferences of the owners / administrators of the repeater you are linking to.

Signals Tab

The **Signals** tab allows you to configure audible announcements that indicate when a variety of system events occur. For each event listed, you can either accept EchoLink's defaults, or create your own custom .wav files. Pressing the speaker button allows you to hear the default announcements or any custom announcements that you may have recorded.

The audible signals defined on this tab are separately settable from the Single-User signals found on the **Signals** tab of the **Preferences** menu.

For those of you in foreign countries, there is also a provision for sending a repeater access tone burst. For those of us in the US, the **Tone Burst** feature is generally set to **Never**.

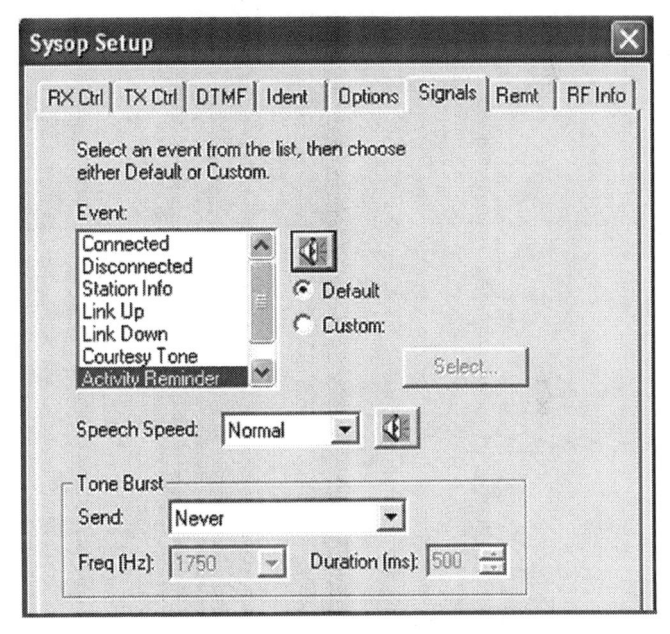

Sysop Event Sounds Setup Window

Remt Tab

The Remote tab allows you to configure two different methods for remotely controlling your station. Since FCC rules require that a station be able to be shut down in case of a problem, you will most likely want to implement one or both of these options. Perhaps the easiest to implement, is **Web remote control**, though you might run into firewall and hostname issues that will need to be overcome. While **dial-in remote control** provides the flexibility to be able to dial-in and use DTMF tones from any regular phone or cell phone, it does require that you interface a TAPI-compatible "voice modem" to the PC. These are special modems that allow you to place and accept voice calls on a PC. You also probably need a dedicated phone line if you opt for the voice modem approach.

When enabled, web remote control allows you to enable and disable the link and perform other basic functions by using any computer's Internet web browser. For more information, refer to the **Remote Control Web Server** section later in this chapter.

Refer to the EchoLink Help file for additional information before buying a voice modem.

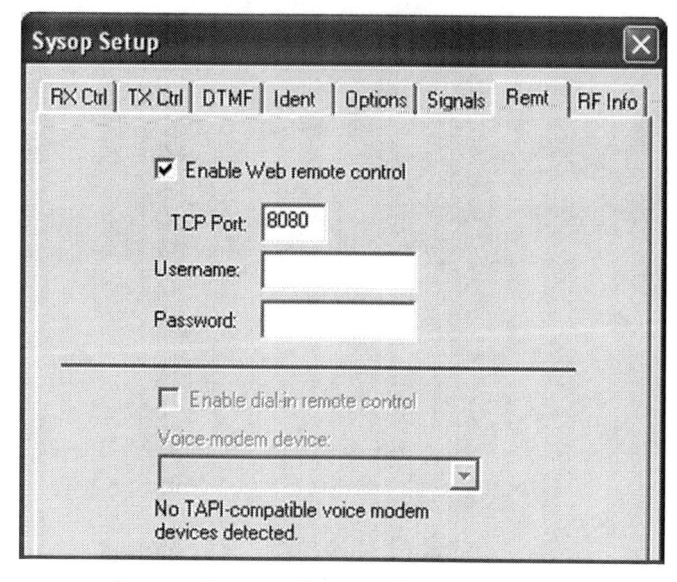

Sysop Remote Control Setup Window

RF Info Tab

If desired, the **RF Info** tab can be used to publicize the location and other details about your node. Upon startup, this information is transmitted to EchoLink's central database on the Internet where it becomes available for display for node searches using the **Link Status** feature found on the EchoLink web page.

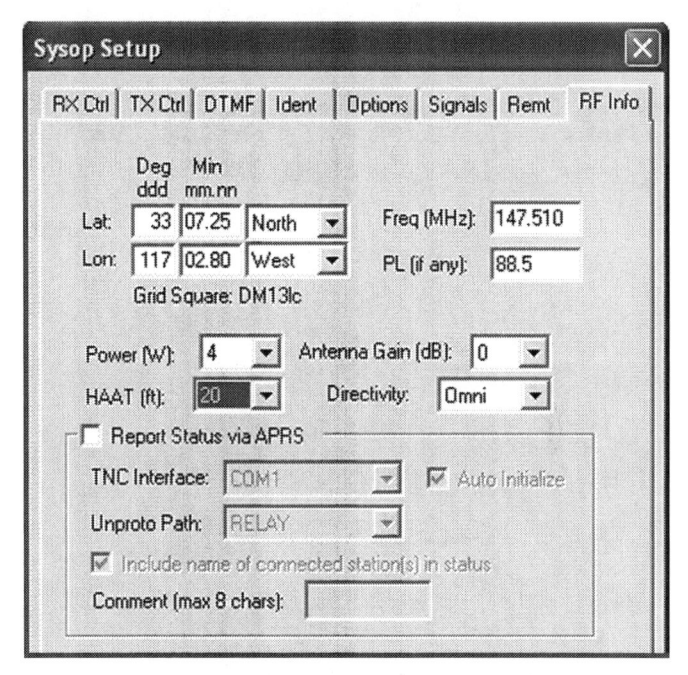

Sysop RF Node Information Window

In general it is desirable to provide your node's **Lat / Lon** coordinates. This is useful to mobile operators looking for nodes they might be able to use, and also to node operators for use as a tool in assessing if near-by stations might be affected when installing a new node. If you prefer not to provide **Lat / Lon** information, enter 00 00.00 for **Lat** and 000 00.00 for **Lon**. Or an approximation of your location can be given without revealing its exact location by rounding off the latitude and longitude.

If you have a TNC connected to your computer reporting APRS information, you may wish to enable the **Report Status via APRS** feature, which allows you to send your station information to local APRS users.

Preferences Connections Tab

When operating an EchoLink node you need to decide if you are going to allow conferences – multiple stations logging into your node for round-table discussions, etc. If you allow conferences there are several related parameters that also need to be set.

Under **Tools**, select the **Preferences** option and then when the **Preferences** window opens, select the **Connections** tab.

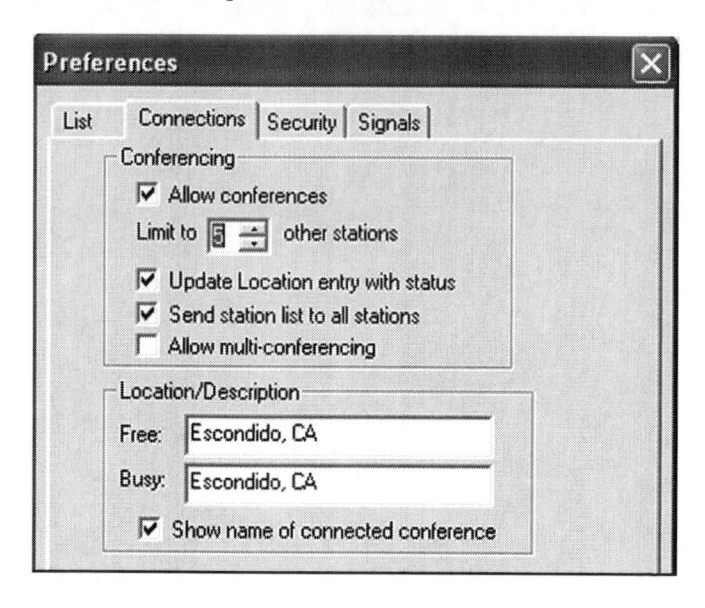

```
Preferences
  List  Connections  Security  Signals
    Conferencing
      ☑ Allow conferences
      Limit to [5]  other stations
      ☑ Update Location entry with status
      ☑ Send station list to all stations
      ☐ Allow multi-conferencing
    Location/Description
      Free:  Escondido, CA
      Busy:  Escondido, CA
      ☑ Show name of connected conference
```

If you choose to **Allow Conferences** you need to determine how many your Internet connection can support. With a dial-up connection, you can only support one connection at a time and conferencing should be disabled. If you have broadband DSL or cable connection you should be able to support up to 7 or 8 connections. With faster Internet connections you should be able to support proportionately greater numbers of connected nodes.

The **Limit to x other stations** sets the maximum number of EchoLink nodes that you will permit to be connected to your node. EchoLink will allow up to the maximum number of connections specified, after which your node will show as Busy, preventing other nodes form connecting.

When other stations view the Station List, selecting **Update Location entry with status** enables them to see how many connections you allow and how many are already connected. These are the numbers seen in brackets at the end the **Location / Description** field in the Station List. This should normally be enabled.

Enabling **Send station list to all stations** instructs EchoLink to periodically update the "text screen" of each connected station with a list of all stations connected in the conference. An arrow highlights the station currently speaking. This should normally be enabled.

In general, **Allow multiconferencing** should be disabled. Checking this box allows one or more other conferences to be connected to your node. If you wish to allow multiconferencing, please refer to the **Conferencing** section of the EchoLink help files to review the problems and special conditions imposed upon this mode of operation.

The two **Location/Description** boxes (**Free** and **Busy**) specify the text that is displayed following your call sign in the Station List. Normally these two fields are set the same, but for added flexibility, the descriptions could be different.

Enabling **Show name of connected conference** allows EchoLink to update the description showing after your call sign in the Station List to display the name of conference servers that you connect to. When in a conference, the "**In Conference XXXXX**" text is displayed instead of your Location/Description as described in the **Busy** box.

Remote Control Web Server

Providing it has been enabled on the **Remt** tab, remote control of your node can be performed using the web browser of any Internet connected computer. Entering the hostname (or IP address) of your link into your web browser will establish communication with your node's EchoLink application, bringing up the remote control web server screen.

Assuming you have enabled the web server on **TCP Port** 8080 using the **Remt** tab, entering **http://your-links-hostname:8080/** into the computer's web browser brings up the following screen.

```
EchoLink - N6FN-L
Time: Thu Nov 18 18:27:58 2010 UTC
[Refresh]

Status                              Text View
Current state: Idle
Link Enabled
Last connection: *ECHOTEST*


Control Link Menu

  ☑ Link Enabled
  ☐ Receive Only
No stations are currently connected.
  ○ Send ID Now
  ○ Disconnect
  ○ Connect to: [        ]
  [ Submit ]
```

As shown above, the **Status** portion of the web server screen provides a brief description of the current condition of the node. The lower section of the screen allows you to enable or disable the node. You can also cause the node to transmit its ID and connect and disconnect to other EchoLink nodes.

The **TCP Port** number, as entered on the **Remt** tab, defines the port on which your node's web server will operate. The default is 8080. You may need to change this setting, however, if your firewall or Internet service provider does not permit inbound traffic on port 8080.

When operating EchoLink behind a firewall (such as a cable or DSL router), be sure to configure the firewall to forward inbound traffic on TCP port 8080 to the computer running EchoLink.

If you are fortunate enough to have a static *IP address* you should have little difficulty getting this to work. On the other hand, if your computer or router is using DHCP (dynamically allocated IP addresses) your IP address may change each time you reconnect to the Internet.

Because of its benefits in configuring networks, it's extremely common to find DHCP, dynamically assigned addressing, being used by your computer, router and Internet Service Provider. The downside for applications like EchoLink's Web Server is that you cannot rely on using the same IP address. It's likely to change from time-to-time. Some Internet providers can provide *static IP addresses*, though it usually costs a bit more.

If you find you are stuck with using DHCP, there is a solution. Instead of using a numerical IP address to access the link computer, by using *Dynamic DNS*, available for free at **http://www.dyndns.com/**, you can assign an arbitrary host name to your node. *Dynamic DNS* also has an automatic update mechanism, which allows a fixed host name to work with dynamic IP addressing. Refer to the *Dynamic DNS* **How To** link on the above web page for full details.

Note: Your IP address and host name can be displayed by right clicking on your call sign in the Station List, and selecting **Show Info**.

If you have taken care of the DHCP problem and are still having difficulty accessing the Web Server from a remotely located computer, refer to the **Firewall / Router Problems** section of **Chapter 4**, which provides some background information on configuring firewalls and routers and links to web sites that might be of further assistance.

Chapter 10: **Eliminating IDs and Tones**

It is preferable to prevent local repeater station identifications and courtesy tone transmissions from being sent out over the Internet to distant nodes. There are several reasons for this:

- When operating on a conference server, significant interruption to the flow of the conversation can occur if station identifications and other noises from the local repeaters of multiple participants are rebroadcast to all participants.

- Re-broadcasting station identifications originating in one country to a repeater located in a different country may be a violation of the rules in that country.

Of course, this is not an issue when implementing a simplex EchoLink node. EchoLink software is the controller in this case. When EchoLink transmits an ID, it only goes out over the link transceiver and not out over the Internet.

Node Co-location at the Repeater Site

If at all possible, it's best to co-locate an EchoLink repeater node at the repeater site. Besides eliminating the need for a link transceiver, detecting signals directly at the repeater's receiver neatly solves the problem of eliminating repeater controller generated IDs and other noises. Since the repeater's controller generates these signals, they only appear on the repeater's output signal.

Lack of suitable Internet availability often prevents co-locating an EchoLink node at the repeater site. In general, a dial-up Internet connection will not be adequate. Something else is required, perhaps DSL or cable. Another alternative would be satellite Internet service, available from many providers across the country.

When a node is co-located at the repeater site, the best way of detecting when a signal is being received is by making use of the COS signal, either directly from the receiver or from an equivalent signal at the repeater controller. Additionally, the audio should be picked up from the receiver, not the repeater's output. Using the receiver's COS and audio connections ensures that EchoLink only transmits to the Internet while a signal is being received on the input and that nothing is added to the transmission.

Another advantage of detecting received signals at the repeater site is a reduction of turnaround delays in the conversation. Instead of using EchoLink's VOX, the **Sysop Rx Ctrl** screen will be set for hardwire detection of the carrier, allowing the **VOX Delay** to be reduced from 1200ms (or more) to around 100ms or less. Additionally, the **Anti-Thump** delay may be able to be reduced from its default of 500ms. Saving a second or more on the turnaround delay significantly improves the flow of conversations through the EchoLink network.

As a side note, when a node is co-located at the repeater site, the EchoLink node will most likely be identified with the call sign of the host repeater with the –**R** suffix attached. When that is the case, there is no need for the EchoLink application to generate ID transmissions for the repeater, thus the **Identify** options located on the **Ident** tab of the **Sysop Setup** screen would not be enabled. Repeater identification requirements are being handled by the repeater's controller. EchoLink connection / disconnection announcements are controlled on the **Options** tab of the **Sysop Setup** screen.

Remote EchoLink Repeater Node Location

When Internet is not available at the repeater site, the link-transceiver, computer and the interface hardware can be located somewhere else within range of the repeater that has suitable Internet access. There are some drawbacks to this setup, but one benefit is that long trips to a mountaintop or other location can be avoided if anything needs to done to the node.

Ideally the link transceiver's location will allow it to receive full-scale, or at least very strong, repeater signals. This improves the node's ability to detect DTMF control commands and provides improved immunity to signal interference.

In addition to eliminating unwanted repeater IDs and tones, an additional issue can arise when using a remotely located transceiver.

If the repeater's hang time (the amount of time the repeater remains in transmission after the input signal has fallen) is too long, it causes EchoLink to keep transmitting until after the tail drops. Repeater hang times in excess of a second or so adversely affect the turnaround time and flow of the conversation. Sometimes it's possible to shorten the duration of the hang time by making an adjustment to the repeater controller's settings.

By using EchoLink's **VOX** threshold level adjustment found under the voice-level meter, its accompanying **VOX Delay**, and enabling the **Squelch Crash Anti-Trip** feature, repeater IDs and courtesy tones can be prevented from going to the Internet. While this is not 100% reliable, with some repeaters it can work surprisingly well.

For detailed information for making these adjustments, refer to the *Setting EchoLink's Received Audio VOX Level* procedures in Chapter 7. If using a general-purpose PC-to-transceiver interface, refer to the instructions found under the *RIGblaster Pro II* section of Chapter 7. If using either the *ULI* or *SignaLink* interfaces, refer to the instructions found in those sections.

Additional information can be found under the **Repeater Linking Tips**, **Carrier Detect** and **RX Control Tab** sections of the EchoLink help files.

Using CTCSS

Another possible solution for these problems is to modify the repeater controller to only output (encode) a CTCSS tone while a signal is being received on the input, thus eliminating the CTCSS tone from the repeater ID and courtesy tone transmissions. If the remotely located link transceiver has CTCSS tone squelch enabled, then it will then only receive (un-squelch) the input's audio signals, ignoring repeater ID and courtesy tones.

A potential problem to using the CTCSS technique is that the repeater controller may not have settings that allow it to selectively apply the tone. If not, on some repeater controllers it might be possible to modify the circuitry (cut circuit trace and jumper style) to achieve the same result.

A potential disadvantage of selectively applying a CTCSS tone to voice signals, is that users of the repeater that have CTCSS squelch enabled on their transceivers will be unable to hear repeater ID's and courtesy tones. This problem can be resolved by informing users of the repeater to not enable CTCSS RX squelch on their receivers. Most amateur radio transceivers have the capability of selectively enabling CTCSS TX encode and CTCSS RX decode.

Passing CTCSS through the Repeater

Some repeater controllers can pass a received CTCSS tone to the output of the repeater. CTCSS tone frequencies are less than 300 Hz, and many repeater controllers (or the receiver) filter out all audio below 300 Hz in frequency, thus eliminating any incoming CTCSS tone. For the method we are describing here to work, the repeater should not filter out audio below 300 Hz, thus allowing the received CTCSS tone to be re-transmitted on its output. This capability might be more common on older repeater controllers, but perhaps some newer ones could be modified to eliminate the low-pass filtering.

The idea here is for the repeater to not output (encode) any CTCSS tone of its own, only passing through any tone present on the received signal. Given that the repeater controller is not encoding a tone of its own, only CTCSS tones present on signals received at the repeater's input will appear at the output. This neatly solves the problem of eliminating repeater generated ID's, courtesy tones and long tails from going to the Internet, since the CTCSS tone will not be present on those signals. Of course, repeater ID's that occur while a person is speaking, will still be transmitted along with the voice.

When a repeater can be setup to operate this way, repeater users wishing to access EchoLink would enable CTCSS encode on their transceivers. And, if they leave CTCSS squelch decode disabled they will still be able to hear the locally generated repeater ID's and courtesy tones. Of course, the link transceiver needs to have CTCSS decode enabled for its receiver, so that it passes only signals with a matching CTCSS code to the EchoLink system.

Preventing Simplex Node Interference

Protecting simplex nodes from being affected by unwanted signals is simpler then when establishing a repeater node. Enabling CTCSS encode and decode on your node will go a long ways towards protecting conversations of hams using the frequency from inadvertently going out over EchoLink. When operating simplex, most hams are not using CTCSS. Consequently if you set up your node with a CTCSS tone not normally used in your area, it is unlikely that inadvertent simplex traffic will interfere with your node's operation.

Chapter 11: **FCC Rules and VoIP**

Because there is a range of opinions as to how the FCC's Part 97 rules apply to VoIP operations, I hesitate to write this chapter. The rules governing amateur radio don't regulate VoIP, IRLP or EchoLink, or the specifics of any other system of operation for that matter. Instead the rules regulate the technical and operational emissions from our transmitters. In general, the rules governing the operation of amateur radio stations, which as licensed operators we are already familiar with, are also applicable when operating EchoLink or IRLP. The FCC's Part 97 rules do not apply to the Internet based portions of VoIP operations; they only come into play regarding the operation of a station's transmitter.

I am not an expert at interpreting the FCC's rules. What follows is an attempt to summarize how the rules may affect amateur radio VoIP operations. The key to understanding how the rules apply is in determining the station classification of RF simplex nodes and repeater linking nodes. The problem is that VoIP linking stations are a bit different than traditional auxiliary or repeater stations, and the FCC's definition of an auxiliary station is rather broad. Yet, it's generally agreed that RF simplex nodes and repeater link nodes are not themselves repeaters. VoIP nodes appear to be a hybrid station type, not neatly fitting into the established categories, but perhaps best resembling an auxiliary station. Since as a licensed amateur radio operator it is your responsibility to make sure your station's operations are in compliance with the rules, you should read them and interpret them yourself.

Rules in other countries are most likely different than those imposed by the FCC on amateur radio operations in the US and its territories. Those operating in foreign countries should consult their own local rules and determine how they may apply to VoIP operation.

Part 97 Definitions

The following definitions are from Part 97, Section 97.3.

(6) **Automatic control.** The use of devices and procedures for control of a station when it is transmitting so that compliance with the FCC Rules is achieved without the control operator being present at a control point.

(7) **Auxiliary station.** An amateur station, other than in a message forwarding system, that is transmitting communications point-to-point within a system of cooperating amateur stations.

(13) **Control operator.** An amateur operator designated by the licensee of a station to be responsible for the transmissions from that station to assure compliance with the FCC Rules.

(14) **Control point.** The location at which the control operator function is performed.

(30) **Local control.** The use of a control operator who directly manipulates the operating adjustments in the station to achieve compliance with the FCC Rules.

(38) **Remote control.** The use of a control operator who indirectly manipulates the operating adjustments in the station through a control link to achieve compliance with the FCC Rules.

(39) **Repeater.** An amateur station that simultaneously retransmits the transmission of another amateur station on a different channel or channels.

Single-User Computer or Transceiver operation

When using your computer or VHF/UHF transceiver to make contacts over EchoLink nodes things are fairly simple. As long as you operate in compliance with accepted practices for making simplex and repeater-based contacts, you are unlikely to experience any difficulty remaining in compliance with the FCC's regulations. If you adhere to the rules that normally govern amateur radio operation, and your transceiver is working properly within power limitation, ID properly, and refrain from commercial traffic, broadcasting, profanity, etc. you should have no trouble staying in compliance.

Establishing Your Own RF Link

When it comes to establishing your own RF simplex and repeater links, you assume the responsibility of being the *control operator*, the amateur that is responsible for the transmissions from that station and that ensures compliance with the FCC's rules. Of course, from an operational standpoint, as the licensee of the station you can share or assign this responsibility to other qualified and willing amateur radio operators.

Assuming that your node installation meets the technical standards for being on a legal frequency, within bandwidth restrictions, lack of spurious emissions, power limitations, etc., we are primarily concerned about being able to shut the link down if something goes wrong. This could be due to some failure in the equipment, or because you need to prevent unauthorized or illegal transmissions from taking place, perhaps because of inappropriate language or subject matter.

The rules require that each amateur station have at least one *control point* from which it can be shut down. In actual practice, for reasons of redundancy and *remote control* you might have two or more.

The rules further require that while a station not authorized to use *automatic control* is transmitting, a *control operator* must be present at a *control point*. Here is where we run into the problem with their being some ambiguity as to what class of station EchoLink belongs to under the FCC's rules. Repeaters are authorized for automatic control, but an RF EchoLink node is not generally considered to be a repeater. If an RF EchoLink node is considered to be an *auxiliary station*, then under the rules *automatic control* is authorized, and a *control operator* need not be present at all times while a station is transmitting. Note that it is now legal for *auxiliary stations* to operate on most 2-meter frequencies.

Remote Control Methods

There are multiple ways of giving *control operators* the capability of shutting down the transmitter from a *control point*.

1. Local Control: If a *control operator* is at a *control point* in physical proximity to the RF link transceiver and computer (*local control*), he could simply turn off the transceiver or shut down the EchoLink software. When not in physical proximity of the equipment the rules require some method of *remote control* sufficient to perform the duties of the *control operator*.

2. Radio Control: In EchoLink's Sysop DTMF settings window, there is a provision for establishing your own DTMF sequences for bringing the **LinkUp** or taking the **LinkDown**. If these commands originate on the simplex or repeater link frequency, they may not work if someone is jamming the frequency or if the link transceiver is stuck transmitting. For the commands to work under those conditions, they need to arrive from some other source, perhaps from an *auxiliary station* on a different frequency. This will generally require some kind of audio mixer to combine control tones received from the *auxiliary station* with the audio being received on the link frequency. As of October of 2006, the FCC now allows the *remote control* function (*control tones*) to occur on 2-meters as well as on the 220 and 440 MHz bands.

3. Wired Link Control: The rules also allow sending control signals over the Internet or a telephone line as acceptable methods of *remote control*. In the Sysop settings, EchoLink software has two methods for providing "wire line" based control support. One method allows the station to be controlled via the Internet from a remotely connected computer or handheld device. A second method requires connecting a "voice modem" to the PC that is running the link's EchoLink software. Using a "voice modem" allows the link to be controlled from DTMF tones generated by any touch-tone telephone. For security reasons both of these methods can be password protected. Control by wire is enabled using the Sysop **Remt** tab screen.

Control Operators and the Control Function

In addition to yourself, you may want to authorize other individuals to act as *control operators* should you not be available to perform that function. There are a variety of reasons you might not be able to perform the *control operator* function in a timely manner: on a trip, at work, sleeping, etc. If the link is running, there should be a *control operator* monitoring so that it can be shut if off if necessary. In addition to being amateur radio operators, persons assigned to be control operators should be familiar with your system and know how to shut it down.

Since control operators will most likely not be present where the equipment is actually operating, you will need to implement one or more of the *remote control* methods previously described. In general, since most hams have the capability of using their radios to send DTMF link and unlink commands, this might be your preferred method of control. Depending upon how you set up your system, these could be on the link frequency itself or via an auxiliary frequency, or both. It's common for a number of individuals to be authorized to perform this function.

As a back up, you might also want to enable EchoLink's ability to be controlled from the Internet. That way, it could still be shut down in case the link is not responding to DTMF commands issued from a transceiver. Since the EchoLink software provides for this capability, and by definition nodes are already connected to the Internet, this option can easily be implemented.

Link Transceiver Call Sign ID

A link transceiver, whether used as a simplex node or a repeater link, requires a call sign for identification purposes. Like any other amateur radio transmission, the link transceiver will have to ID using its assigned call sign (the link owner's most likely) at least once every ten minutes during a communication and at the completion of a communication. The ID can be performed by phone or Morse code. If performed by Morse code, it must be at a rate not to exceed 20 wpm. See FCC 97.119 Station Identification.

Auxiliary Control Station Frequencies

If using an *auxiliary* station to implement a *control point*, Section 97.201, of the FCC's rules states:
  "(b) An auxiliary station may transmit only on the 2 m and shorter wavelength bands, except the 144.0-144.5 MHz, 145.8-146.0 MHz, 219-222 MHz, 222.00-222.15 MHz, 431-433 MHz, and 435-438 MHz segments."

FCC Part 97 Rules and Regulations

If you don't have a copy of FCC Part 97, W5YI maintains it on line at: **http://www.w5yi.org**   Click on the **FCC Rules Part 97** link found at the left hand side of the page.

Part 97 is also available on the ARRL's web site:
http://www.arrl.org/part-97-amateur-radio

Chapter 12: **Simplex Link Frequencies**

If you are setting up your own RF simplex link node, you will need to select a non-interfering simplex frequency. Just staying off the simplex FM National Calling Frequencies is not enough. Depending upon the band and region of the country you are in, this may not be as easy as taking a quick look at the recommended ARRL band plan. In some of the more urbanized parts of the country finding a non-interfering simplex frequency might be somewhat of a challenge.

While the published ARRL band plan is a good place to start, you really need to look into the recommendations of the "recognized" frequency coordination committee in your area. In many states the frequency coordination committee has established a set of linking frequencies that are to be used. If that is the case in your area, you need to contact the coordinator and verify the frequency you should be operating on.

If you don't know who that is or how to contact them, the following URL to the SERA (South Eastern Repeater Association) web site provides links to frequency coordination councils all over the United States and Canada. SERA's band coordination plans are quite clear, and in the absence of other information may provide some guidance.
http://www.sera.org/links.html

You might also check with whoever administers your local club's repeater; they will certainly know how to contact the frequency coordinator in your area.

Appendix A: **PC Serial Port Pinout**

For reference, the following diagram shows the 9-pin RS-232 serial port configuration found on most PCs and USB-to-serial adapters.

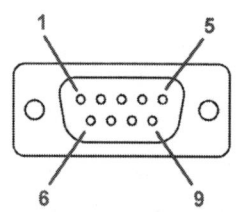

View is into the PC's Connector

Note:
When wiring from back of the connector, pinout is mirror image of that shown above. (Pins 5 and 9 are to your left, 1 and 6 to the right.)

PC Serial Port Pinout			
DE-9 Pin	Name	Signal Direction Relative to PC	Description
1	CD	←	Carrier Detect
2	RXD	←	Receive Data
3	TXD	→	Transmit Data
4	DTR	→	Data Terminal Ready
5	GND		System Ground
6	DSR	←	Data Set Ready
7	RTS	→	Request to Send
8	CTS	←	Clear to Send
9	RI	←	Ring Indicator

When a signal direction is shown away from the PC's connector, it indicates that circuitry within the PC is driving the signal. Conversely, when the direction is shown pointing to the PC, circuitry within the PC is receiving the signal.

Note: PC's often have similar shaped connectors with 3 rows of pins, these are not serial port connectors, it's a video connector.

Appendix B: **Entering Call Signs via DTMF**

Two of EchoLink's default DTMF commands (Connect-by-Call and Query-by-Call) require the entry of the station's alphanumeric call sign. Using a radio's DTMF keypad for entering alphanumeric characters requires pressing two keys for each letter or number:

- **Letters**: The first key pressed is the key on which the letter appears, and the second key pressed is a **1**, **2**, or **3** indicating which of the letters on that key is being entered.
- **Numbers**: To enter a number, press the desired number key followed by 0 (zero). Press zero twice to enter a zero.

1 QZ	2 ABC	3 DEF
4 GHI	5 JKL	6 MNO
7 PRS	8 TUV	9 WXY
*	0	#

For example, to enter the following alphanumeric characters:
- **E**   Press key digits **3, 2**
- **Y**   Press key digits **9, 3**
- **Z**   Press key digits **1, 2**   (see Note 1)
- **4**   Press key digits **4, 0**
- **0** zero   Press key digits **0, 0**

Note 1: The EchoLink DTMF keypad, as illustrated above, may be labeled different than the one shown on your radio. Notice that the letters **Q** and **Z** appear under the **1** key. Many keypads show **Q** under **7**, and **Z** under **9**, but that is not how character entry is performed in EchoLink.

Note 2: Callsigns need not be entered in full. If a partial call sign is entered, EchoLink locates the first match under stations currently logged on. If a match is not found, EchoLink will announce, "NOT FOUND".

Appendix C: **Radio Setup Guides**

Nifty Ham Accessories produces laminated guides for almost all Kenwood, Icom and Yaesu radios sold since the year 2000.

Nifty set up and programming guides are available from most ham radio retailers and also directly from Nifty Ham Accessories.

**Nifty! Quick Reference Guides
Available for all recent Model
Kenwood, Icom and Yaesu
Transceivers**

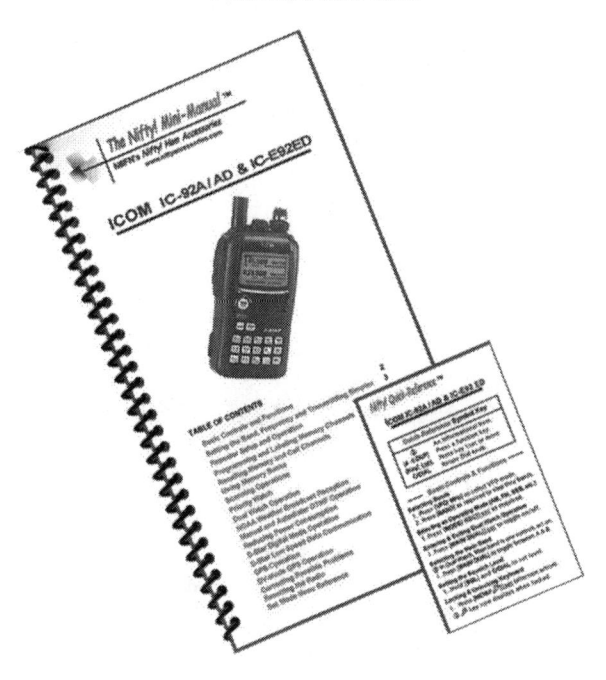

Condensed and simplified step-by-step operating instructions for all menus and modes of operation.

Kenwood

Transceiver Reference Guides

TH-K2AT / K4AT	TM-271A
TH-F6A / F7E	TS-480HX /SAT
TH-D7A(G)	TS-570 D/S(G)
TM-V7A	TM-D700A
TH-22 / 42AT	TM-G707A
TS-50S	TM-D710A
TM-V71A	TM-V708A
TH-G71A	TS-870S
TM-261 / 461A	TS-2000 / 2000(X)

Nifty! Ham Accessories

760-443-0281   www.niftyaccessories.com

Icom
Transceiver Reference Guides

IC-T2H	IC-80AD	IC-718
IC-T7H	IC-91A / AD	ID-880H
IC-Q7A	IC-92A / AD	IC-746PRO
IC-P7A	IC-207H	IC-756PRO
IC-V8	IC-208H	IC-756PRO(II)
IC-R20	IC-910H	IC-756PRO(III)
IC-T22 / 42A	IC-2100H	IC-910H
IC-W32A	IC-2200H	IC-7000
IC-T70A	IC-2720H	IC-7200
IC-T81A	IC-2820H	IC-7600
IC-V80	IC-703 (P)	IC-7700
IC-V85	IC-706MKII	IC-7800
IC-T90A	IC-706MKII(G)	IC-V8000

Nifty! Ham Accessories

760-443-0281 www.niftyaccessories.com

Yaesu
Transceiver Reference Guides

VX-1R	FT-50R	FT-1000MP MKV
VX-2R	FT-60R	FT-1000MP Field
VX-3R	FT-90R	FT-2000 / FT-2000D
VX-5R	FT-100D	FT-1802M
VX-6R	FT-250R	FT-1900R
VX-7R	FT-270R	FT-2600M
VX-8R	FT-450	FT-2800M
VX-8DR	FT-817 / 817ND	FT-2900R
VX-8GR	FT-847	FT-7100M
VX-120	FT-857 / 857D	FT-7800R
VX-127	FT-897 / 897D	FT-7900R
VX-150	FT-920	FT-8800R
VX-170 / -177	FT-950	FT-8900R
VR-500	FT-1500M	FTM-10R
		FTM-350R

Nifty! Ham Accessories
760-443-0281 www.niftyaccessories.com

Made in the USA
Charleston, SC
19 March 2012